Otherworldly Antarctica

OTHERWORLDLY

Ice, Rock, and Wind
at the Polar Extreme

Edmund Stump

WITH ILLUSTRATIONS BY Marlene Hill Donnelly

The University of Chicago Press :: Chicago and London

ANTARCTICA

The University of Chicago Press, Chicago 60637
The University of Chicago Press, Ltd., London
© 2024 by Edmund Stump
Illustrations © 2024 by Marlene Hill Donnelly
Source of basemap data for Map 2: http://lima.usgs.gov, courtesy of the US
Geological Survey
Published 2024
Printed in China

33 32 31 30 29 28 27 26 25 24   1 2 3 4 5

ISBN-13: 978-0-226-82990-6 (cloth)
ISBN-13: 978-0-226-82991-3 (e-book)
DOI: https://doi.org/10.7208/chicago/9780226829913.001.0001

Publication of this book is made possible by a generous grant from the
Alfred P. Sloan Foundation.

Library of Congress Cataloging-in-Publication Data

Names: Stump, Edmund, author. | Donnelly, Marlene Hill, illustrator.
Title: Otherworldly Antarctica : ice, rock, and wind at the Polar extreme /
    Edmund Stump ; with illustrations by Marlene Hill Donnelly
Description: Chicago : The University of Chicago Press, 2024.
Identifiers: LCCN 2023037503 | ISBN 9780226829906 (cloth) | ISBN
    9780226829913 (ebook)
Subjects: LCSH: Antarctica—Description and travel. | Antarctica—Pictorial
    works.
Classification: LCC G860 .S85 2024 | DDC 559.89—dc23/eng/20231016
LC record available at https://lccn.loc.gov/2023037503

♾ This paper meets the requirements of ANSI/NISO Z39.48-1992
(Permanence of Paper).

This book is dedicated to my children,

Simon, Molly, and Nick,

and to children everywhere.

You are the meaning of life.

Contents

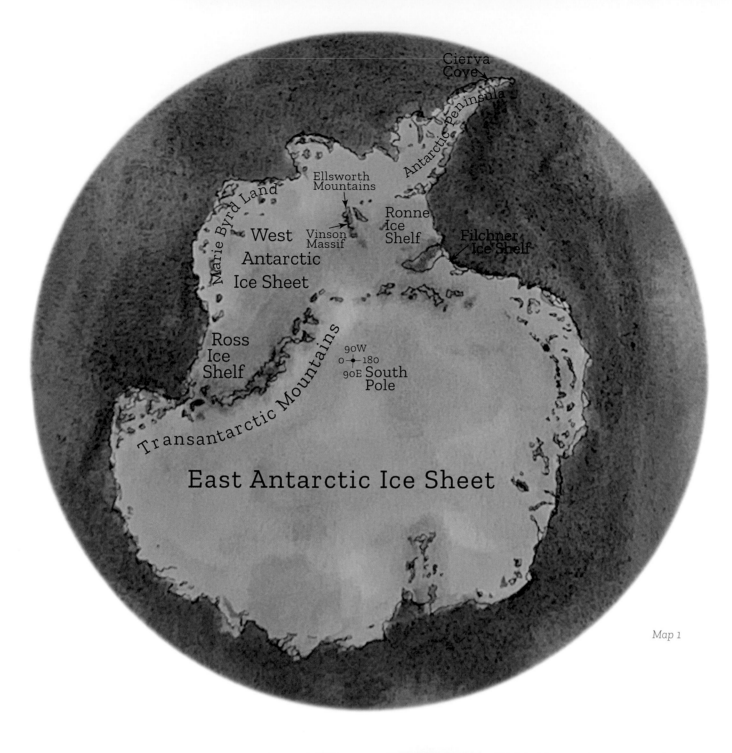

Cierva
Cove

Antarctic Peninsula

Ellsworth
Mountains

Marie Byrd Land

West
Antarctic
Ice Sheet

Vinson
Massif

Ronne
Ice
Shelf

Filchner
Ice Shelf

Ross
Ice
Shelf

Transantarctic Mountains

90W
0 — 180
90E South
Pole

East Antarctic Ice Sheet

Map 1

Gothic Mountains

Mount Erebus, the world's southernmost active volcano

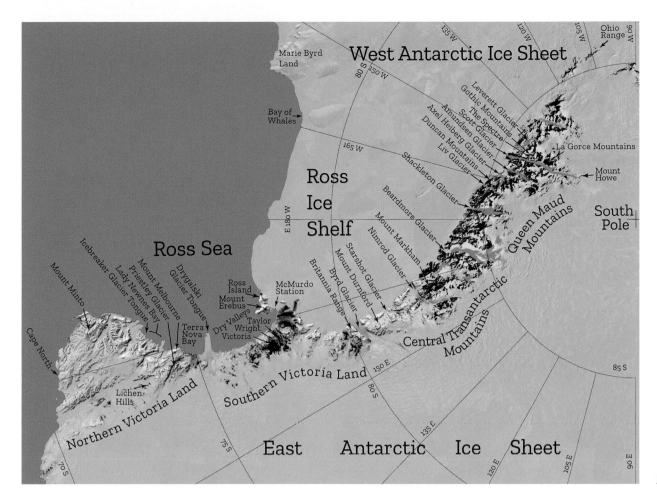

West Antarctic Ice Sheet

Marie Byrd Land

Bay of Whales

Ross Ice Shelf

Ross Sea

Ohio Range

Leverett Glacier
Gothic Mountains
The Spectre
Scott Glacier
Amundsen Glacier
Axel Heiberg Glacier
Duncan Mountains
Liv Glacier

La Gorce Mountains

Mount Howe

Shackleton Glacier

Beardmore Glacier

Queen Maud Mountains

South Pole

Mount Markham

Nimrod Glacier

Starshot Glacier

Mount Dunaford

Byrd Glacier

Britannia Range

Mount Minto

Icebreaker Glacier Tongue

Mount Melbourne
Priestley Glacier
Lady Newnes Bay

Drygalski Glacier Tongue

Terra Nova Bay

McMurdo Station

Ross Island
Mount Erebus

Dry Valleys
Taylor
Wright
Victoria

Cape North

Lichen Hills

Northern Victoria Land

Southern Victoria Land

Central Transantarctic Mountains

East Antarctic Ice Sheet

Map 2

Greetings from the summit of the Spectre

Drygalski Glacier Tongue

Liv Glacier

Preface

Fortune smiled in 1970 when I landed a position on a research project bound for the Transantarctic Mountains of Antarctica. As an undergraduate geology major—and passionate outdoorsman—I had dreamed of doing research in big, rugged mountains. The western United States certainly qualified, as did any number of far-flung ranges around the world. I knew that a mountain range existed in Antarctica separating the East and West Antarctic Ice Sheets. But that was it. I had no idea about the geology and even less about the otherworldly landscape of ice and rock that awaited me.

The project was funded by the National Science Foundation's Office of Polar Programs through the Institute of Polar Studies at the Ohio State University. The project leader was David Elliot, a rising star in Antarctic geology, and he brought with him a dozen other researchers, including faculty and graduate students, each assigned a major group of rocks in the area. The paleontologists in the group were searching for

vertebrate fossils following successes the previous season: the first bones of an Antarctic *Lystrosaurus*, the Triassic mammal-like reptile found on all the other Southern Hemisphere continents, had been discovered by Elliot and his team. My assignment was to collect and analyze the rocks in the lower levels of the mountains, which had formed during a mountain-building episode 500 million years before. I was psyched.

The field party worked out of a pair of remote field camps in the Queen Maud Mountains, 500 miles south of coastal McMurdo Station on the tip of Ross Island, supported by three Huey helicopters of the Navy squadron VXE-6. Every day after breakfast, weather permitting, the pilots would set us out in pairs to climb around in the mountains, mapping and collecting samples, before picking us up and bringing us back to camp in time for dinner.

The first of the two camps was in the middle of McGregor Glacier. The glacier filled a valley that was surrounded on three sides by steep snow walls topped with outcrops of rock. About five miles to the east of camp McGregor Glacier joined Shackleton Glacier, the outlet glacier that flows through that stretch of the Transantarctic Mountains. Downstream from their confluence, Shackleton Glacier runs due north for 45 miles through a canyon cut 5,000 feet deep into the rock.

For the first week, I mostly mapped along this corridor, both at glacier level and high on ridgelines of the surrounding mountains, then worked out from there. The second half of the season we moved camp 120 miles to the southeast and continued the routine of daily helicopter support. Because of how much fuel was needed to safely return to camp, the helicopters were limited to a 100-mile radius, and I pushed that limit. By the end of the season, I had mapped and sampled a 300-mile stretch of the Transantarctic Mountains.

The logistics were incredible. I was chauffeured anywhere I wanted to go by Navy pilots who had honed their skills in Vietnam and relished the challenges of mountain flying. Sometimes the places they landed me were high and edgy, barely wide enough to put down both helicopter skids. But even if the landing site was a piece of cake, we always reached it by flying over spectacular, alien terrain.

Each passing day as I studied the rocks, I became more aware of the ice, its pervasiveness, its innumerable forms, its grand scale, its incredible detail: the knobby textures of ablation-pitted glacier ice, the ubiquitous *sastrugi* or wind-carved patterns in snow, bubbles and cracks in meltwater ponds, the swirls in ice-cored moraines, windswept drifts hung on ridgelines, shattered seracs in surging icefalls, orderly crevasse fields and chaos, foreshortened distance on rising surfaces of snow, and the vast ice shelf as flat as the sea upon which it floats. Although my business in Antarctica was the study of rocks, it was the all-pervasive ice that took hold on me.

From that first encounter with Antarctica, I wanted to share what I was experiencing. Nature at the polar extreme was like nothing I'd ever seen or imagined. The landscape was stark and utterly pristine, the vistas vast with not a sign of life anywhere in that empty world. Over the years I managed to study and photograph all the major segments of the Transantarctic Mountains. As a result, my photo collection is uniquely comprehensive.

I feel profoundly privileged to have had the opportunity to work in Antarctica. During my forty-year career, I logged thirteen field seasons mapping and collecting throughout the 1,500-mile length of the Transantarctic Mountains as well as once in the Ellsworth Mountains of West Antarctica. For several seasons I worked out of large, helicopter-supported camps, but the way I liked it best was four souls at a remote location in the mountains, on our own with our gear—snowmobiles, sleds, tents, food, and fuel.

After retiring from teaching for thirty-seven years at Arizona State University, I served as the Global Perspectives Guest Speaker on a Lindblad/NatGeo Expeditions cruise to the Antarctic Peninsula in December 2014, where I had the opportunity to witness another part of Antarctica. The peninsula—the "Banana Belt" to those of us who have worked in the Antarctic interior—is a dramatic maritime landscape of ice, rock, and open water which I happily added to my Antarctic photo collection.

Since the National Science Foundation is a federal agency, I must thank the American taxpayers for having supported my marvelous career. As with any project funded by NSF, mine were judged worthy by peer review, and the payback was publication in the scientific literature. More information about my publications and the research personnel on these expeditions is available in this book's appendix.

Most often journal publications are of little interest beyond the scientific community for whom they are written. This book is for everyone. It is my homage to Antarctica, the continent of ice, and to the Transantarctic Mountains, my stomping ground.

THE SPECTRE AND THE GLORY

I have a craving for solitude, carried forth from my youth, that feeling of being alone with Nature, not another soul within miles. As a kid, my favorite vista was from the crest of the cemetery, set on a hill above my hometown and commanding a sunset view of the valley, where the Juniata River emerged from the narrows and spread a broad floodplain between Appalachian ridges.

When the prospect of Antarctica first loomed, I imagined that I had found the ultimate refuge for solitude. But in reality, for safety reasons, one never goes out alone in the field; your partner is always there. And although the place you've gotten to may transcend, and you both feel it, still you're not alone. Several times I had hiked the hills up behind McMurdo Station to a place on the backside beyond the hum of motors, where I was alone with McMurdo Sound and a hundred miles of the Transantarctic Mountains. Nevertheless, the station *was* just over the hill, and the aura of the largest

metropolis on the continent imbrued the edge of the scene. Another chance for solitude was in the field after dinner, while the others were typically in a deep, digestive mode. Then I would go out and fool around in the vicinity of camp. If camped on snow, there was always the possibility of sastrugi to focus my meditation, or if on moraine or blue ice, the patterns that arose there.

But there was that one, memorable day beneath Mount Durnford, the sun, the warmth, the rising fog. I was out of sight and earshot, seriously alone, when a spectre appeared to me in all its glory. The second half of the 1978–79 season, I was working in the area south of Byrd Glacier. A helicopter placed my four-man party at a series of sites where we would camp for several days, map the geology, and then move on. This site was a bowl on the northeast side of Mount Durnford (7,136 feet), one of a series of peaks that populate the crestline of the Churchill Mountains. We had camped on snow next to a ridge of bedrock that in one direction rose 4,000 feet along a steep spur directly to the summit and, in the other, played out for a mile or so at a nearly horizontal level. The day before, we had climbed the spur, mapping the folded limestone as we went. The ridgeline cast jagged shadows across the smooth, undulating lower slopes. As I caught my breath, I also caught this image.

2

A few hundred feet below the summit, we reached the level where our mapping stopped. But with so little distance left to go, we pushed on to the top. The Churchill Mountains run for nearly 150 miles in a north–south direction between Byrd and Nimrod Glaciers. They divide the East Antarctic Ice Sheet directly to the west from the rising foothills of the Transantarctic Mountains to the east. Mount Durnford projects prominently to the east of the main summit line, giving a commanding panorama of the entire range both to the north and to the south.

When we reached the summit, the view was expansive, but what really set the scene with foreground interest was a thumb of cross-bedded sandstone marked with a most unusual, crinkled, weathering pattern. I said, "Boys, if you want to be famous, crawl over there and I'll take your picture." And so they did: grad students Pat Lowry and Scott Borg, and field assistant/mountaineer Phil Colbert. Those dirt-bags were no posers, but they posed for me that afternoon.

The day in question dawned, still and cloudless. A helicopter was scheduled to move the camp, but several hundred feet below and as far as the eye could see a ground fog spread beneath the cobalt sky. Since helo pilots don't fly over terrain they can't see, the party was on its own till the fog cleared. Everyone was feeling lazy after the exertion of the previous day, so we sat around the cook tent making pancakes and reading books.

After lunch I asked if anyone wanted to walk down the ridge from camp, secretly hoping that they all would say no. Indeed, each did demur, so I set off alone for an afternoon of solitude. The sun was to the west, skimming the face of Mount Durnford, part shadow, part light. The fog radiated brilliant white, with no hint of detail or texture. Only at its edges where it lapped up onto the mountain could I see wisps of vapor, barely undulating as the fog crept higher. The air did not move. I soon was carrying my parka on my arm and still was warm.

The rocks along the ridge were limestones, like those on the face of Durnford the day before, but these were less broken and preserved sedimentary textures indicating possible formation in a tidal flat. I reminisced about a field trip to the Bahamas while still in grad school, two little outboard-motor boats putting up a tidal channel at low tide, walking barefoot out across the pure, white mud of the drained lagoon to fringes of desiccating scale. In my mind, I stepped from that mudflat on Andros to this Early Cambrian tidal flat at my feet, 520 million years in the past. The fog bank became the sea. It was deep and teaming with new life forms. But none had yet evolved to land. I was alone on that rocky shore, strolling along. The sun was intense. I took off my shirt and T-shirt.

Backlit and peering down into the murkiness, I noticed the fuzzy shadow cast by the ridgeline above and behind me. Then the hair on my arms stood up, but not from the cold. Deep in the fog at a spot on the edge of the shadow, a circle of light appeared. When I moved down the ridgeline, it followed. In a quarter hour, the circle had grown more intense with distinctly rainbow colors, and it centered on a dark spot that could only be my shadow.

I waved my arms widely and could barely make them out in the figure at the center of the light. At times I had used the term "aura" as a figure of speech to describe someone's or something's essence, but I had never seen one, and I didn't believe that others really had either. Nevertheless, I could see that the colors were encircling me, or at least my projection. So, I was the progenitor. But what made the rainbow? It reminded me of the interference patterns seen through a petrographic microscope, where convergent, polarized light is refracted as it passes through a thin section of rock.

I walked another half mile down the ridge, my aura in tow. As the sun moved further around in the sky, its rays grazed the face of an ancient drift at the foot of Mount Durnford, illuminating its complexion with knife-edged sharpness. I shot some photos.

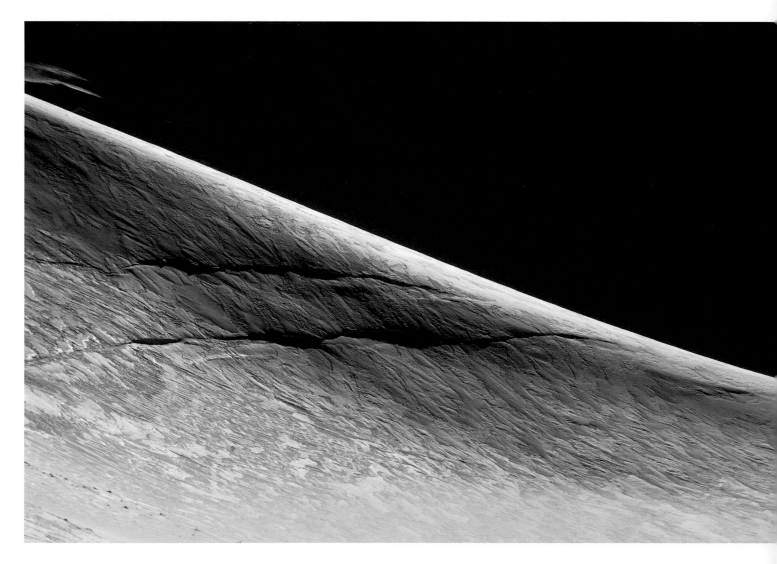

The air was dead calm. The brilliant fog crept silently up the landscape. I stripped off the rest of my clothes, except for sunglasses, and carefully lay down on my outstretched parka. The funk of two weeks of hard exercise emanated into the air space. The heat of the sun was intense, and I imagined the UVs zapping the microflora and fauna that had taken up residency on my skin during the previous fortnight. I closed my eyes and drifted away.

Maybe a half hour later, a slight chilling of the air brought me to consciousness. The fog had risen over the ridgeline and was beginning to reflect sun rays back into the sky. The aura had gone missing. It was time to amble back to camp. I took my time, enjoying the thickening atmosphere as Durnford disappeared and the brilliance enveloped me.

The atmospheric phenomena that I saw that day were new to me, but not to others. My shadow was a Brocken spectre, and the rainbow, a glory. The spectre is a shadow cast downward by an observer into cloud or mist. It requires that you be in sunshine above clouds, and it helps if the angle of the sun above the horizon is close to the slope of the terrain. The name derives from the Brocken, the highest peak in the Harz Mountains of Germany, long associated with witchcraft and where Walpurgis Night fires are still lit to this day. Clouds shroud the mountain most of the year, and its domed summit affords proper slopes in all directions to produce the phenomenon.

The glory is a rainbow-colored circle in cloud or mist that occurs around the point directly opposite the sun from the observer. In practice, this is the head of the observer's shadow. Its name derives from the halos, or glories, that surrounded saints' heads in medieval art. The classic explanation of the glory is that light is diffracted as it interacts with droplets of water vapor in the cloud, but the actual path of the process is still debated.

THE ICE

Antarctica is a continent brimful of ice that spreads outward from the interior and spills into the Southern Ocean (see Map 1). The East Antarctic Ice Sheet reaches a thickness of two miles and is so weighty that when it formed, starting ~45 million years ago, it pushed down the interior of the continent to depths below sea level. The smaller West Antarctic Ice Sheet is grounded on continental crust that is naturally beneath sea level. Confined by Marie Byrd Land and the base of the Antarctic Peninsula on one side and the Transantarctic Mountains on the other, its outward flow creates giant floating ice shelves, the Filchner, the Ronne, and the Ross. The Ross Ice Shelf is the size of Texas, with an average thickness of 1,300 feet and an ice cliff at its terminus that rises 50 to 150 feet out of the water.

Dividing East and West Antarctica, the Transantarctic Mountains rise majestically to elevations above 14,000 feet and extend for 1,500 miles into the interior, where at the Ohio Range the East and West Antarctic Ice Sheets overtop the mountains and merge. The mountains act as a dam to the outflow of the East Antarctic Ice Sheet, which backs up to elevations in excess of 8,000 feet before spilling through in a series of mighty outlet glaciers that furrow deep, broad valleys across the mountains and drain into either the Ross Sea or the Ross Ice Shelf.

Outlet Glaciers

In the austral summer of 1908–9, in their bid to be the first to reach the South Pole, Ernest Shackleton's party discovered Beardmore Glacier, a 100-mile-long passage through the mountains. The party followed the snowy, less-crevassed western (right) mar-gin of the glacier, passing the eastern (left) side of Buckley Island, avoiding the heavily crevassed middle of the glacier. The Beard-more swells to a width of 30 miles where Mill Glacier joins it at the tip of the Domin-ion Range (left rear of image).

Downstream from Mill Glacier, Keltie Glacier (left) merges with the mighty Beardmore at the northern tip of the Supporters Range. The distance from the rock to the bottom of the image is approximately four miles.

The sketch shows a time during the last glacial maximum, circa 25,000 years ago, when half of the Britannia Range was under the ice that rose up its side due to the thickening and grounding of the Ross Ice Shelf and its subsequent backup into the Byrd drainage. The rock in the foreground of the preceding photo was completely overridden by the glacier.

Fifteen miles across Byrd Glacier, the 10,000-foot south face of the Britannia Range is hung with surging glaciers and icefalls. The Byrd flows from left to right in glacier-time at a rate of about six feet per day. With the largest catchment of any of the outlet glaciers, Byrd Glacier delivers 24 billion tons of ice to the Ross Ice Shelf each year.

In its lower reaches Byrd Glacier floats on the water of the Ross Sea. In its upper reaches it slides on rock. The grounding line, where the glacier begins to float, is abreast of this view. The faceted, bare-rock walls on the far side of the glacier attest to a time when Byrd Glacier was grounded for its entire length, with the result that the glacier thickened and rose higher in its valley, cutting the faces adjacent to the glacier today.

For scale, a surreal herd of giraffes wanders through the maze of ice. At 19 feet tall, even the biggest male cannot look over the walls of this fragmented wasteland.

In a 10-mile stretch midway down Nimrod Glacier, the flow constricts from a width of 15 miles to 8 miles, producing a cascade of shattered ice. Shot from a helicopter at low altitude into the maelstrom, this image reveals an otherworldly terrain of ice in chaos. For scale, the numerous vertical faces in shadow are about 30 feet high.

1
3

With its sensuous curves, islands in the stream, crevasses ruffling broad swaths of the surface, and an arcing pressure ridge in its lower reaches, Liv Glacier is the quintessential outlet glacier.

As with Scott and Shackleton in Great Britain, Amundsen in Norway, and Mawson in Australia, Richard E. Byrd is synonymous with Antarctic exploration in the United States. He led his first expedition in 1928–30; it was based at the Bay of Whales on the eastern edge of the Ross Ice Shelf and featured extensive airborne exploration for the first time. He chose Liv Glacier as his gateway to the polar plateau and successfully navigated this 40-mile-long corridor on the first flight to the South Pole.

I crossed the lower reaches of Liv Glacier with a snowmobile in 1974.

Eighty miles long and four to six miles wide, Amundsen Glacier is steep and heavily crevassed in its central portion. A ribbon of textured blue in the center rear of the image, it flows from left to right through a multitude of ridges. The flat-topped escarpment in the distance bounds the polar plateau.

Tier upon tier of ragged peaks line the banks of Scott Glacier as it wends its way for 130 miles across the mountains, dropping 9,730 feet from the first crevasses at the edge of the polar plateau to sea level, where the glacier enters the Ross Ice Shelf.

CHAPTER 1

Crevasses

Ice is the strangest substance, a crystalline solid that floats. Move it slowly and it flows like taffy, move it fast and it shatters like glass. In this image from the upper Beardmore Glacier drainage, blue ice flows slowly and smoothly across a shelf of bedrock, but upstream where the valleys are steep and confined, the ice surges down crevasse-riddled icefalls and breaks into ragged ice blocks or *seracs*.

Propelled by gravity, glaciers flow down valleys like mighty rivers, although they do it in glacier-time—feet per day, miles per year. Unlike water, however, which is a liquid, ice is a solid composed of crystals. The movement of ice in the solid state is called *ductile flow*. At the boundaries of the ice crystals, water molecules frantically migrate (or *diffuse*) from points of higher to lower stress, causing the ice crystals to shift their shapes. In glacier-time, Kent Glacier courses its way down the east side of the Markham Plateau.

When the speed at which ice deforms reaches a certain threshold, no matter how fast the ice crystals are rearranging at the molecular level, they cannot keep up with the increasing pull of gravity and so the ice fractures. Glacier ice is most brittle at the surface and becomes increasing ductile at depth due to its own overlying weight, so cracks originate at the surface and propagate downward. Brittle fractures open incrementally, typically separating by not more than hundredths of an inch per break, over time producing gashes in glaciers known as *crevasses*.

In this image, crevasses festoon central Scott Glacier. Grizzly Peak stands out on the left. The summit of the Spectre peeks out at the center.

Crevasses are the hidden danger of Antarctica, where thin bridges of snow may span deep openings, gravity wells that accept whatever falls in, ready to devour man or machine. In areas where blowing snow accumulates and crevassing is active, each brittle pop opens the gash a bit and snow trickles down the crack, filling it in. As the crevasse continues to open, the crack fills only so far before the incoming trickle of snow clogs the opening and freezes, in time creating a bridge of hard snow over a widening and deepening empty space. Bridges may collapse, opening their depths to view, or remain in place covering the chasm beneath. The depth to which a crevasse can open is approximately 150 feet. Below that, ductile flow reigns, and cracks cannot penetrate more deeply. The notion of a "bottomless" crevasse is myth.

Descend into the otherworld of deepening blue.

CHAPTER 1

Whether broad and arcuate or short and
frazzled, crevasses reflect the interplay of
gravity, brittle failure, and terrain.

2
1

Ice bursts radially from a bulge on the south face of the Britannia Range. The otherworld explodes.

Twin icefalls lap the 2,000-foot face of Horney Bluff, reaching toward the raging margin of Byrd Glacier, stuck in avalanche cones of their own creation.

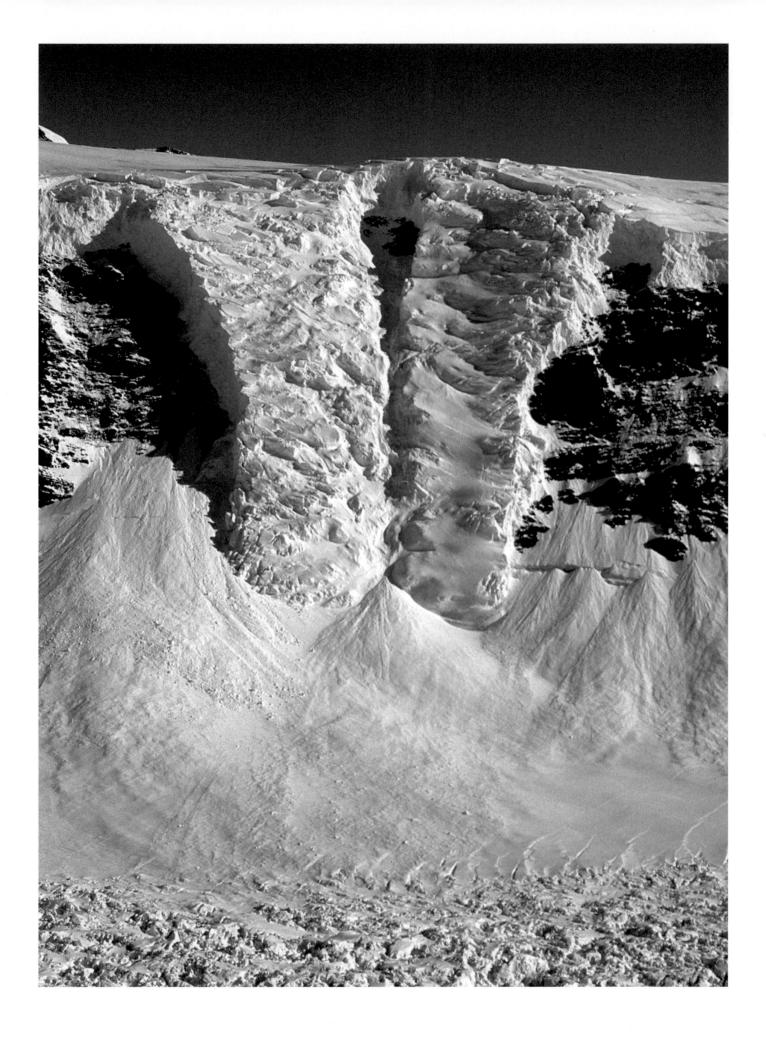

Scott Glacier begins its descent from the East Antarctic Ice Sheet at Mount Howe, the southernmost outcrop of rock on the planet, pictured on the horizon 80 miles to the south. With a width of seven miles, the glacier flows past the La Gorce Mountains on the left and Mount Gardiner on the right. Extremely straight and uniform crevasses in the middle of the glacier give way to ragged scars on the margins.

Crossing Liv Glacier

My first direct encounter with an outlet glacier was the Liv, which we crossed with a single snowmobile and sled on December 26, 1974. We had no air photos for the route, only the verbal account of a New Zealand geologist who had blazed the crossing a decade

earlier. It was a maze of crevasses. We were green and had little clue of what we were getting ourselves into.

The biggest barrier to crossing lower Liv Glacier is a pressure ridge that runs down its middle (see photo on page 14). This spine of ice heaves up as much as 10 feet on its sides and splits down the crest with a deep set of cracks. On the near side of the ridge was a depression in the glacier that was filled with snow, making a sort of super-highway along the upturned barrier. If we followed that south to the point where the pressure ridge ended, we could round it and head straight across the glacier after that.

Visibility was excellent as we headed out. The surface was low-relief sastrugi that were fairly soft, taking about a two-inch track from the snowmobile. It said to me that

2
4

CHAPTER 1

the sastrugi were also fairly fresh, probably laid down during our previous storm, so it might have covered some otherwise apparent crevasses. This put me on edge, which is always good for holding one's focus. I imagined myself seventy feet down in a crevasse, wedged so tightly in the narrowing walls that I couldn't move and barely breathe, feeling the ice rapidly draining the heat from my body. Driving as slowly as I could without stalling, scanning the surface in front of me, I approached the pressure ridge.

I had read the NSF's helpful advice in its manual *Survival in Antarctica*. However, the first sentence in the "Crevasse Crossing" section read: "Experience reveals that there is greater danger of serious injury . . . when a person is in a vehicle than when he or she is skiing or walking." The manual did not elaborate on how the vehicle was supposed to proceed without a driver.

Some parties in the early years of snowmobiling in Antarctica attempted to follow their machines on skis while steering with a pair of lines attached to the handlebars. Another line was attached to a kill switch to stop the engine, and typically the thumb throttle on the handlebar would be taped in an open position. Word was that the setup was extremely cumbersome to manipulate. Inevitably, someone lost control and fell, dropped the lines, missed the trip switch, and the snowmobile putted into the distance to disappear over the horizon.

The notion of remote-control snowmobiling seemed impractical to me, plus I didn't know how to ski. My plan, in the event the Ski-Doo went down, was to levitate. Rather than relaxing in the seat, I sat with my legs cocked under me, so that at the first hint of the bottom dropping out, I would spring into the air and land on terra firma. Happily, I can say that I never needed to execute this maneuver.

The pressure ridge rose up like a giant serpent, twisting down the axis of the glacier. Along its ablation-pitted back, blue ice sparkled like scales in the sunlight. In the photo, Phil Colbert and my advisor, Charlie Corbató, are checking our map. The pressure ridge crosses from the right. The faint ruffled snow surface above the figures' heads is an obvious crevasse field. The superhighway that we followed is in between. Mount Ferguson (3,900 feet) is the dark peak (far right) five miles across Liv Glacier. The Prince Olav Mountains with elevations over 13,000 feet mark the skyline 25 miles to the south.

The sketch shows Byrd's flight path up Liv Glacier in 1929 and the track that I followed in 1974. My route led from the end of the Duncan Mountains on the left across smooth snow to the pressure ridge in the middle. From there it turned up the glacier to the end of the cleft, then turned right and traveled straight across the glacier to the far side.

25

I started driving slowly up the passage, intent on the smooth surface of white. Several hundred feet along I drove onto the edge of a bridged crevasse, noticing the hairline crack at the last moment and braking with about two feet of the snowmobile's nose ski hanging over the lip.

The crevasse opened clean for about 50 feet down, but it was narrow enough to jump across, and with the base of our snowmobile eight feet in length and the Nansen sled even longer, I knew we could drive straight across, no problem. I revved the throttle. Because of a tight tow rope, the machine leaned back into the opening with its track whining and snow flying, then jumped forward, the sled in tow, and we were on to the next. We encountered another dozen or so crevasses along the corridor, but found each to be similarly narrow, and we drove across.

Crossing crevassed terrain deals in trade-offs. A person's foot exerts considerably more pressure per square inch than does a snowmobile or sled, so a thin, narrow bridge might break through and eat you,

where a machine would drive across with the bridge remaining intact. Alternatively, a person might walk across a thick, widely bridged crevasse with impunity, where the weight of the machine would cause the span to collapse. Crevasses have you coming and going.

When we reached the end of the pressure ridge, we rounded the serpent's tail and headed toward the closest rise on the other side. After a mile or so I spotted a thin crack not far ahead, off to my right. I stopped and probed it. It looked okay so we crossed. But other cracks appeared and soon Phil was earning his meager wage as he plunged his ice axe two or three feet into the hard snow. If it didn't break through into space, he would take one step forward and do it again. At first, he found several crevasses capable of eating the snowmobile, so we had to work up and down their lengths to find a spot thick enough to cross. After a half mile of the exercise, Phil wasn't breaking through, and we resumed traversing with the snowmobile in the lead, heading straight toward the far bank of Liv Glacier.

Several hundred yards short of the margin I turned and drove toward the mouth of the glacier. Rounding the toe of the Tusk, a spectacular horn of pure marble, we continued to the south side of Mount Henson where we pitched camp on hard snow for our two-night stay. We had learned a lot that day about the wiles of a glacier. We had been close to it, probing, choosing our way, completing the crossing. In hindsight this was the most dangerous day of glacier travel in my career. Sometimes the gods smile on the brashness of youth.

Sea Ice

Every winter the Southern Ocean surrounding Antarctica freezes over, and every summer it breaks up and melts back into the sea from which it came. In the span of a year, one tilted circumnavigation of the sun, sea ice progresses from the insubstantial to the substantial and back again, through phased lengthening and shortening of days and nights, and with that the melting warmth of our star and the icy cold of the firmament.

As temperatures plunge, the first tiny platelets of ice begin to form, creating a thickening sludge called *frazil*. If the sea is calm, this will set into a stiff surface layer of ice. If not, currents and winds break the frazil into pieces, with their edges upturned from the splash of impact with their blunted neighbors, so-called *pancake ice*.

In this image, freshly formed frazil follows the shoreline of Hut Point, at a spot where seasonal ice pulled away from the shore and frazil started to fill the breach. Motion fractured the nascent ice and water oozed up along the cracks forming the lips, but the fragments had not separated, as is typical of pancake ice. The frazil was 4–6 inches thick.

The icebreaker in the distance is bound for McMurdo Station just beyond the point. Vince's Cross figures the crest of Hut Point. At the southern end of McMurdo Sound 35 miles distant, Mount Discovery stands sentry with a fata morgana rising at its base.

When warming temperatures return in the spring, the carapace of sea ice softens and begins to break. Long cracks, called *lead*s, fracture the shell, cutting pathways through an intricate mosaic of last winter's freeze.

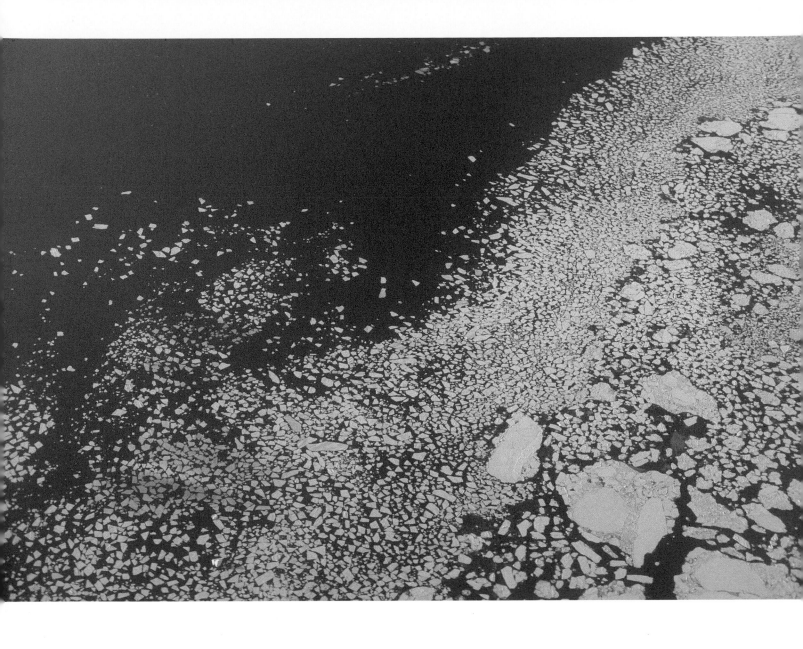

As sea ice fragments, it forms polygo-
nal pieces called *floes*, which collectively are
known as the *pack*. The ice pack drifts with
the currents and is driven by the wind. This
image was shot from the port of a Hercules
C-130 at around 25,000 feet.

Cutting paths in light pack in the lower center of the image, Zodiacs ferry tourists to a landing in Orne Harbour. These 25-foot inflatable rubber boats propelled by outboard motors are the workhorses between ship and shore in the Antarctic Peninsula.

Midsummer storm clouds gather over
Gerlache Strait. Sea ice almost gone.

Icebergs

When the great ice sheets of East and West Antarctica reach the margins of the continent, they calve chunky icebergs into the Southern Ocean. By far the largest icebergs come from the ice shelves, whose thicknesses may be as much as a quarter mile. Over periods of years and tens of years, long cracks penetrate the floating ice shelves, freeing huge, flat-topped, tabular icebergs that can be miles in length. In contrast to an ice shelf's flat-topped bergs, when glaciers calve into the ocean, they produce irregular icebergs in all shapes and sizes. Because of its maritime climate and ice-choked mountains rising steeply from the sea, the Antarctic Peninsula excels in iceberg production.

Within the short period of their existence, icebergs produce some of the most beautiful, transient sculptures of the natural world. These monuments to decay array the periphery of the continent, adrift on driving seas, congregating in sheltered bays, slipping off their outer layers, ultimately succumbing to the rising warmth.

In this early summer photo from November 1981, small icebergs choke the inlet at Cape North, the northernmost point of land in northern Victoria Land. Fed by the small glacier in the background, chunks had calved into open water the previous year but did not drift out of the inlet before the winter freeze.

An enclave of the otherworld exists at Booth Island in the Antarctic Peninsula where icebergs have grounded and formed a maze of icy canyons.

Ablation (both melting and evaporation) takes its toll, and in the process creates icy freeform sculptures.

Mating dance of the icebergs, Cuverville Island.

CHAPTER 1

Cierva Cove

The Zodiac putt-putted softly away from the mother ship into a profound stillness. Our driver cut the engine and invited us to absorb the silence and the scene.

We were suspended on a transparent surface surrounded by bits and pieces of sparkling crystal, tinged blue beneath the waterline. A dome of matte gray illuminated the shadowless landscape. Ringing the cove were steep walls of ice fed by glaciers from the slopes above. Their faces bore the scars of tension and release where they had calved the icebergs that spread throughout the cove. The ice of the walls was young, only faintly blushing blue.

Bare rock reached to peaks through the snowfields above the ramparts and held the wall along one ice-free side of the cove. The sparkling, transparent ice that occupied the bay was a feature known as *brash ice*, formed either by the disintegration of ice calving from an ice wall or by storm breakup of freezing sea ice, in this case probably the former. We drifted in an enclave unto itself, sheltered from currents and storms. Ice was everywhere.

I had been scanning the cove and spotted a strangely vivid iceberg floating low in the distance. When I asked if we could check it out, the driver was happy to oblige. As we approached, I could make out tilted layers of brown within the layered blue. It measured about 300 feet, a football field, in length.

As we drew closer, I began to make out large fractures splaying back into the mass. The nearer we came the deeper I peered into the transparency. Forms billowed, streaks shot through, creatures appeared. Brash ice lapped the bottom of the face.

CHAPTER 1

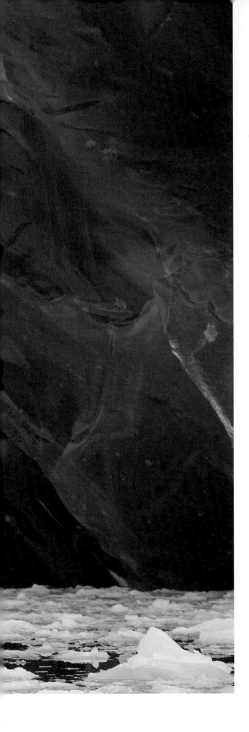

I peered through my camera and the cove and all context vanished. I was in an abstract world of hue and form, zooming in and out, each frame its own distinct reality, Nature caught in the act. My pulse quickened. Horizontal distance, 10 feet.

From there we headed into a portion of the bay where a group of small icebergs had gathered. Brash bits were scattered about, but not in the density we'd seen earlier. The bare-rock wall reflected blackness. We glided past the display with barely a comment, rapt in the scene. One berg had tipped up and its belly fallen out, producing a natural arch to rival any on the Colorado Plateau. Blue shone on the underside of the bridge as if illuminated from within. The sky was the texture of felt. As we drifted past, the axis of the arch aligned with a small, mushroom-shaped berg in the background, and in the deep distance the orange huts of an Argentine base.

This was enough alignments for one morning. The mother ship had slipped quietly up behind us and was now calling her brood of Zodiacs back to the nest. We queued at her side, then disembarked into the changing room. Amen to Cierva Cove!

Glacier Tongues

Along the coastline of the Transantarctic Mountains, glaciers flow, or *debouch*, directly into the Ross Sea creating a series of floating glacier tongues. For most of the year these elongate features are bound in seasonal ice, but following the summer breakout they float free in a sea of Prussian blue. To the north, the glaciers originate in the mountains, whereas to the south they draw their ice from the East Antarctic Ice Sheet. When a glacier leaves the confines of land and enters water, it surges forward in glacier-time and its margins fracture into saw-toothed patterns. The flights between Christchurch, New Zealand, and McMurdo Station follow this dramatic coastline, and I never missed a chance to gawk from the porthole in the door at the back of the plane.

4
5

The coastline of Lady Newnes Bay is an intricate interplay of land heads and glacier tongues that jut into the Ross Sea, ice-covered in this early summer image. The edge of Icebreaker Glacier Tongue holds the foreground. Next comes Parker Glacier Tongue, followed by Aviator Glacier Tongue. Some 50 miles to the south stands the dark cone of Mount Melbourne (8,957 feet), a dormant volcano that rises at the head of Terra Nova Bay.

Petite Parker Glacier Tongue is three miles across and ten miles in length.

Originating in the mountains to the west of Mount Melbourne, Campbell Glacier pokes into the northern end of Terra Nova Bay, forming the Campbell Glacier Tongue.

Priestley Glacier spills from the ice sheet through the cleft in the back and flows off the left of the image.

Surpassing all others in sheer size, Drygalski Glacier Tongue issues into the southern end of Terra Nova Bay where the East Antarctic Ice Sheet pours through a low breach in the mountains. The bright line encircling the Drygalski is a 150-foot-tall wall of ice. The width of the front of the glacier tongue is 15 miles. The length back to the mountain front is approximately 40 miles.

The British National Antarctic Expedition (1901–4), commanded by Robert Falcon Scott, discovered and charted the glacier tongue in January 1902. The ice walls were so high that the lookout was unable to see over them from the crow's nest of HMS *Discovery*, the expedition's ship.

Finger Rafting

When I flew south in late October 1980, I was treated to a rare and remarkable display of Nature's creation. Terra Nova Bay was already clear of the previous winter's ice. In its place thin ice had formed over the dark, open water, and wind or currents had pushed it against the fast ice at the edge of the bay. As a result, the compressed ice had cracked and begun overriding itself, but rather than one side going up and the other down, the up and down directions alternated along the fractures, producing *finger rafting*.

Meltwater Ponds

I could drown in the depths of a meltwater pond, bound in lacy veils. On solstice days when rocks warm just enough, adjacent snow and ice will melt. The meltwater collects in low spots and there it freezes. Ponds typically occur on moraines, those glacier-deposited accumulations of rock debris, but also in depressions on rocky terrain. For example, each of the small volcanic craters that follow the ridgeline up from Hut Point behind McMurdo Station has a meltwater pond in its dimple. The patterns result from an interplay of fracture and the release of dissolved gases as the water freezes.

4
9

If a rock tumbles onto the ice of a meltwater pond, because it absorbs sunlight more efficiently, it may melt the ice beneath it. The result is that the rock burrows into the ice over time and may sink inches below the surface.

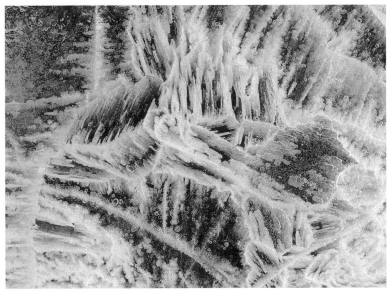

The Ineffable Ice Puddle

December 14, 1974: I awoke that morning to the sound of gurgling water and thought, what is going on? This is 85° south, in the interior of Antarctica. But it was unmistakable. A brook was running just behind camp. When I rolled out of my tent to look, for sure there was a little, fast-moving stream, maybe two feet across and a few inches deep, coursing through the boulders of the moraine. The stream originated off to the west somewhere where the warmth of the sun on this near-solstice day was heating rocks just enough to cause melting of the adjacent glacier ice. The water ran off a couple hundred yards to the east and ponded up on the smooth ice of an established meltwater pond. For anyone journeying in this part of the world, a significant part of each day is taken by firing up the stove and melting snow to supply the party with enough water. So that day we had a boon: we dipped water directly from the stream into our bottles and hit the slopes a half hour earlier than usual.

That evening after dinner, I was out on the moraine enjoying the otherness of the polar landscape. The sun had gone behind the ridgeline to the south, and the melting in the moraine had again frozen. The stream that had been running so vigorously in the morning was gone, its path now a trail of slick, undulating ice. The evening was windless, so I strolled with open parka. The silence was broken only by the soft hiss of the gas stove in the cook tent, and soon enough I had walked out of range even of that.

In the habit of field geologists everywhere, I scanned the ground as I walked along. Out ahead, an ice puddle appeared. At a distance, it looked like something different but nothing special. It was a patch of extremely smooth ice at a point where the morning stream had overflowed its bank and flooded a small, low spot on the moraine, maybe four feet long and a couple of feet across. Up to that time I had seen a number of meltwater ponds, always in their frozen state and always near sea level where temperatures did on rare occasions reach the melting point. They ranged up to a hundred feet across and typically were built of transparent or translucent ice, shot through with bubbles that blurred into gray in the solid, icy depths of the ponds.

But this one was like nothing I'd ever seen. It was a thin, maybe quarter-inch, sheet of ice, delicately suspended at its rocky edges over a space of air. Paper-thin bubbles speckled the ice, which had begun to freeze when the puddle was at its fullest. Then as the stream flow abated, the puddle drained, and as it did, the line where the air bubble advanced on the water along the underside of the ice became diaphanous threads draped across the frosted face. Extra to that, a web of ice crystal boundaries etched by the air underpinned portions of the puddle.

It was one of those moments when you fall down the rabbit hole. I was drawn in, down on my hands and knees, my face as close to the ice as focus would allow. I was aware of nothing else. This was a two-dimensional world, delicate beyond belief, a galaxy of bubbles, intricate and intermingling, contracting, swirling, bursting. It made my head spin.

Prior to the season, I had purchased a set of three macro rings (1x, 2x, 3x)—lenses that attach to the front of a camera lens, like a filter, and enable close-up or macro photography. For the next hour, I screwed them on and off my 50 mm lens, framing portions of the frozen galaxy that had materialized that afternoon and likely would not last another day.

This was the moment when I discovered the otherworld that exists in the lens of a camera. It is a rectangular, two-dimensional world of pattern, form, and detail. It may be vast, it may be wee. It may be strange, it may be beautiful. It is a glimpse of reality and an other reality created by framing the glimpse. Since that day, I have pursued this otherworld whenever I am in Nature.

Wafer-thin ice spans its stony boundaries, spattered by bubbles, beset with sweeping curves as liquid left, then frosted by dry air. Horizontal distance, ~18 inches.

Retreating water left its curvy mark on this portion of the puddle. Horizontal distance, ~12 inches.

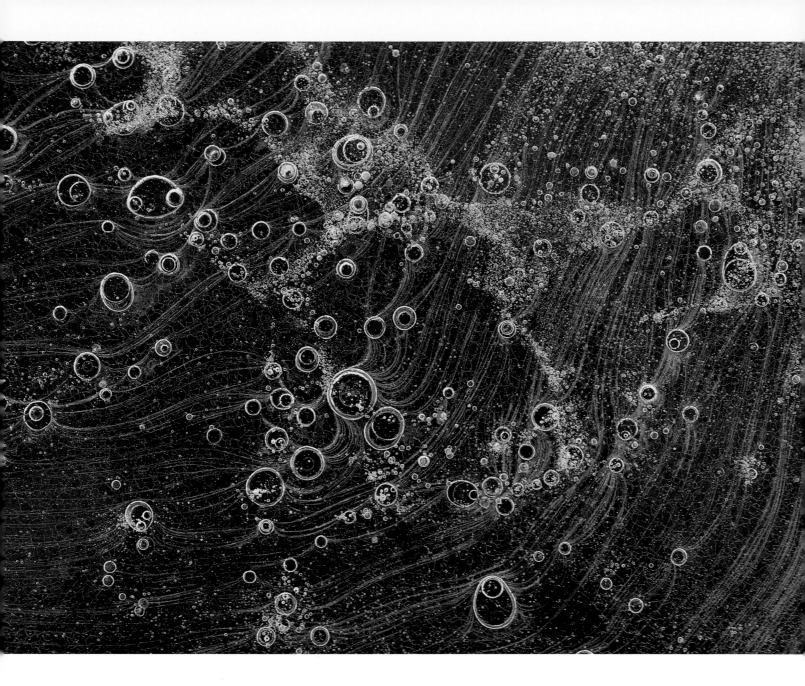

This frozen puddle: a galaxy of bubbles
vapor-etched and frail.

THE ICE

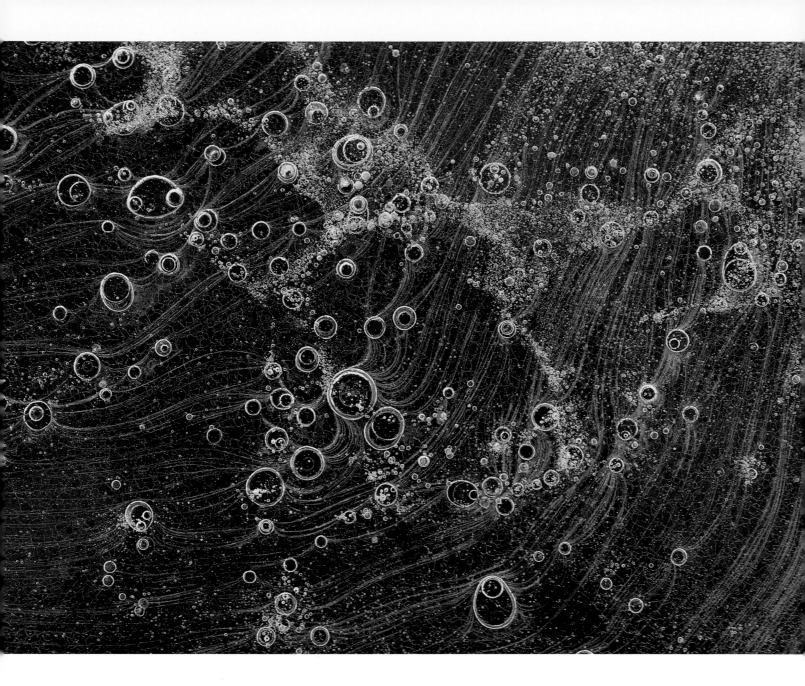

This frozen puddle: a galaxy of bubbles
vapor-etched and frail.

5
5

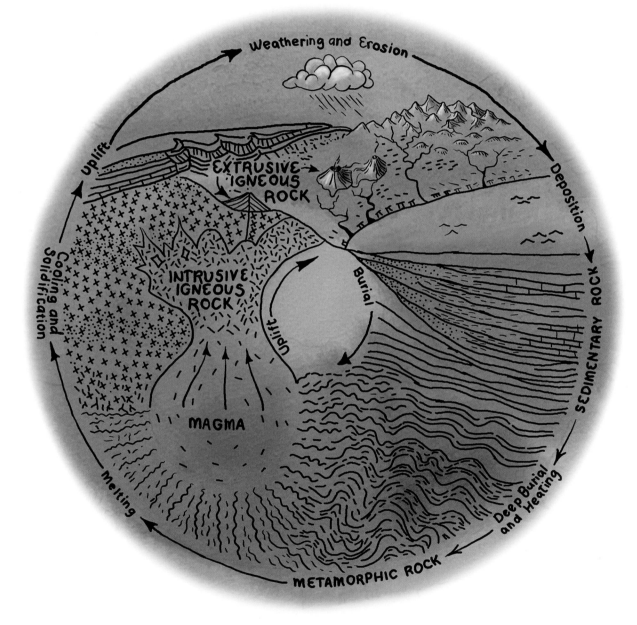

The Rock Cycle, geology's Wheel of Karma

2

THE ROCK

Since Earth's beginning more than 4.6 billion years ago, its rocky crust has evolved through episodes of mountain building, cycles of creation and destruction that played out in rock-time (inches to feet per million years), and have left a wrinkled, splotchy record across its aging face. Antarctica's rocks are no different from rocks found on any of the other continents. They all followed the Rock Cycle, geology's Wheel of Karma, encompassing the life, death, rebirth, and accumulated suffering of rocks on Earth. Each cycle of the wheel is a mountain-building event, distinct in time and place.

All rocks (except for coal and volcanic glass) are composed of minerals, and all minerals are composed of atoms in specific proportions, arranged repetitively in specific crystal structures. Imagine you are deep in the guts of an active mountain belt. The rocks are suffering increasing temperatures and atoms in the mineral lattices are vibrating so energetically that they begin to break free of their bonds and become liquid. Melting progresses and the sticky fluid that results is called *magma*. Magma is less dense than the rock from which it melts, thus it rises buoyantly, pushing aside and consuming the surrounding rock as it ascends into cooler levels of the crust. As the magma cools new minerals begin to grow, and when the last bit of liquid has crystallized to the solid state, an *igneous* rock is born. If the magma crystallizes within the crust it is said to be *intrusive*. If it reaches the surface and erupts, it is said to be *extrusive* or *volcanic*. Because an intrusive rock cools slowly, its crystals grow large enough to be seen. An example is granite, the most common igneous rock in an active mountain belt. Because a volcanic rock cools quickly, crystallization produces numerous, extremely small crystals too small to be seen by the unaided eye. Basalt, which composes the floor of all Earth's ocean basins, is an example.

As the mountain belt rises, weathering and erosion attack and wear away the rocks at the surface. These break down to clay and grains of sand, so-called sediments, or dissolve into water. Currents of water and wind carry them until they come to rest, typically on the flood plains of rivers or in the ocean along the margins of continents. Another common sediment is lime mud, formed in warm, shallow seas not by the breakdown of minerals but by the secretions of organisms, such as plankton, coral, and sea grass. As sediments accumulate, those at the bottom of the pile experience higher temperatures and pressures, transforming them by the processes of compaction and cementation into *sedimentary* rock. Sand becomes sandstone. Clay becomes mudstone or shale. And lime mud becomes limestone.

Greater depth produces even further transformation. Increased temperature and pressure break down some minerals and grow others. This recrystallization is called *metamorphism* and the resulting rocks are *metamorphic*. At even greater temperatures, these metamorphic rocks will melt, thus completing the Rock Cycle.

Most of East Antarctica is made of continental crust formed during mountain-building episodes between 1.4 and 3.4 billion years ago. The Transantarctic Mountains are younger. They initially formed during a mountain-building episode a half billion (500 million) years ago along what was the continental margin of Antarctica at the time.

Erosion reduced those mountains to a level plain. Then about 350 million years ago sediments, sand and mud, began to accumulate on that old erosion surface. Deposition of these sediments ended about 180 million years ago, when huge volumes of basalt magma intruded in sheets (sills) between layers of the sedimentary rocks and erupted volcanically. This relationship of the eroded innards of a mountain belt overlain by horizontal layers of sedimentary and intrusive rocks occurs throughout the Transantarctic Mountains.

Antarctic geologists informally call the older rocks beneath the erosion surface the "Basement" and the overlying sequence of horizontal sedimentary rocks the "Beacon" (short for the formal name, Beacon Supergroup). These names will figure in the passages that follow.

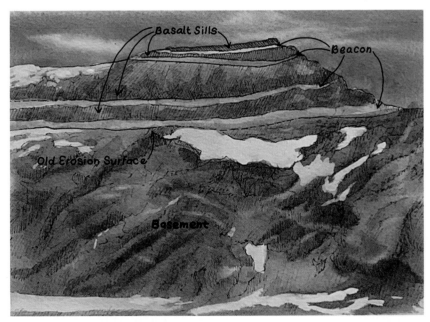

This sketch labels the three major components of the Transantarctic Mountains as displayed in the Lichen Hills: the Beacon, several basalt sills, and, beneath the old erosion surface, the Basement.

This 600-foot face in the Lichen Hills of northern Victoria Land encapsulates the three major episodes of the geologic history of the Transantarctic Mountains. The Basement is the folded and intruded outcrop in the lower portion of the cliff. The light-colored parts are granite, the dark-colored parts, metamorphic rocks. The Beacon is represented by the thin, light-colored layers immediately above the erosion surface and at three horizons in the upper cliffs. These thin layers of sandstone are separated by four distinct sills of basalt that form the dark cliffs.

The box outlines the following photo.

59

At the peak of mountain building, the dark-colored metamorphic rock flowed like taffy in liquid magma. In rock-time, feet per million years.

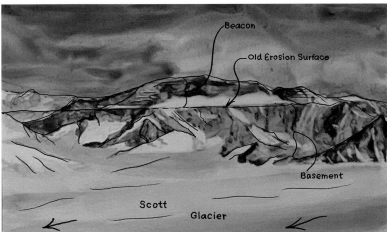

Beacon over Basement, the same geology at Mount Blackburn as at the Lichen Hills

One thousand miles away, another example of Basement overlain by Beacon occurs on the east side of Scott Glacier across the face of Mount Blackburn (10,745 feet). The old erosion surface beneath the layered Beacon is visible on the left side of the image and to the right is marked by the horizontal line at the base of the snowfield.

Transantarctic Mountains

The present-day Transantarctic Mountains began to rise about 45 million years ago. But this uplift was different from typical mountain building, which involves thickening of crust, deformation, melting, and volcanism. The Transantarctic Mountains simply tilted upward as a block with the rocks that compose them remaining unchanged. This occurred when the continent fractured along a fault, with West Antarctica subsiding below sea level and adjacent East Antarctica rising to form the mountains. The trace of this fault forms two graceful arcs along the mountain front, one facing the Ross Sea and the other facing the Ross Ice Shelf, with their intersection at McMurdo Sound (see Map 2). Geologists call such a feature a *rift shoulder*, and the Transantarctic Mountains are the grandest example on Earth.

Perhaps it was no coincidence that glaciation began in Antarctica at about the same time as uplift of the Transantarctic Mountains. Ice has been the consort of these mountains since their beginning. Perhaps it also is no coincidence that the present-day Transantarctic Mountains follow the same orientation as the 500-million-year-old mountain belt that forms their foundation. One could say that's Karma.

In this image, looking northwest along the Queen Maud Mountains, the entire width of the Transantarctic Mountains is displayed. The Ross Ice Shelf marks the right horizon and meets the mountains closer in. The East Antarctic Ice Sheet lies just beyond the edge of the escarpment at the left horizon. Mount Goodale (8,432 ft) centers the image.

Northern Victoria Land

Northern Victoria Land is the northernmost portion of the Transantarctic Mountains. In summer months it has coastline with open water along its northern and southeastern shores. With a breadth of 300 miles, it is twice as wide as any other sector of the mountains. The interior of northern Victoria Land is a vast region of low-profile ranges with ice reaching up to their ridgelines. The terrain culminates at Mount Minto (13,665 feet), the highest point in the region, shown here in the upper right.

The view is toward the east over Evans Ridge in the foreground.

63

Mountains rise dramatically out of the Ross Sea along the southeast coastline of northern Victoria Land. Mount Monteagle (9,121 feet) is the culmination of the ridges in the center of the image. Aviator Glacier Tongue frames the bottom.

Southern Victoria Land

Southern Victoria Land begins at the Drygalski Glacier Tongue in the north and ends at Byrd Glacier in the south. The northern and southern portions of southern Victoria Land are characterized by snow-clad mountains and outlet glaciers, but what distinguishes the region are three valleys, Victoria, Wright, and Taylor, adjacent to McMurdo Sound that once hosted outlet glaciers but today are ice free. Due to an uplift of the mountains in this area or a lowering of the ice sheet, outflow of ice was constricted, the glaciers retreated, and the dark walls of the valleys absorbed the summer sun, maintaining this ice-free oasis on a continent of ice.

6
4

Victoria Valley is the northernmost of the Dry Valleys. Like its counterparts to the south, it has a lake in the middle fed by meltwater from both ends of the valley. This is Lake Vida, viewed up valley. McKelvey Valley separates to the south (left) and Victoria Upper Valley separates to the north (right) around Sponsors Peak, the high ground in the middle. The East Antarctic Ice Sheet reaches to the backside of the cliffs at the skyline.

The stream that feeds Lake Vida from Victoria Upper Glacier in the summer flows through a terrain of *patterned ground*. This feature is due to the annual freeze/thaw cycle that concentrates fine-grained sediment in cells and expels coarse clasts to their boundaries. The cells measure approximately 10 feet across.

During the 1982–83 field season I conducted a reconnaissance survey of the Basement rocks throughout the Dry Valleys, with day trips from McMurdo Station and several tent camps in the valleys. Bull Pass, a saddle between Wright and Victoria Valleys, was the site of one of the camps. Looking down Wright Valley from the pass, three hanging glaciers, Goodspeed, Hart, and Meserve, adorn the smooth valley wall, fed by snowfields atop the Asgard Range. The seasonal Onyx River flows from left to right, continuing in the next figure, where it delivers water to Lake Vanda.

CHAPTER 2

Looking up Wright Valley from the same
vantage as the previous figure, Lake Vanda
nestles between the Asgard Range on the
left and the Olympus Range on the right.
The lake receives meltwater from both ends
of the valley during the peak of summer.

Central Transantarctic Mountains

South of Byrd Glacier the Churchill Mountains run parallel to the coast for 140 miles, gently rising through a broad lowland of ridges and névés (snowfields) and abruptly climbing to a blocky crestline that banks the East Antarctic Ice Sheet. South of Starshot Glacier, the Churchill Mountains split into four separate ranges that terminate at Nimrod Glacier. South of the Nimrod, at the northern end of the Queen Elizabeth Range, Mount Markham (14,270 feet) towers at the culmination of a spectacular ridge system that surrounds its blocky summit on three sides. Ice plasters all surfaces up to the ridgelines where bedrock peeks through. The Markham summit is the peak on the right of the pair (just left of center). On the eastern side, the Frigate Range (right horizon) rises boldly to the summit. Crevasse-riddled glaciers surge down its flank. The summit plateau drops off to the left (south).

In December 1986 a helicopter landed
my party and a snowmobile at 9,000 feet in
a saddle in the Queen Elizabeth Range, 13
miles to the south of Mount Markham. We
drove up the long, gently inclined ramp to
the summit plateau where we mapped the
folded limestones. The view from the summit of Mount Markham was surpassing. The
entire 100-mile length of Nimrod Glacier
was visible from the plateau to the Ross Ice
Shelf. A dramatic ridge dropped from the
summit and splayed north for 25 miles out
to the bank of the Nimrod.

CHAPTER 2

Queen Maud Mountains

During the austral summer of 1911–12, a Norwegian expedition lead by Roald Amundsen challenged Robert Falcon Scott and the British Antarctic Expedition of 1910 in a race to the South Pole. Scott followed Shackleton's route up Beardmore Glacier and reached the pole a month later than Amundsen, perishing on the return in one of the most enduring tragedies of polar exploration. At that time, the mountains and outlet glaciers beyond Beardmore Glacier were unknown. Amundsen's approach from the Bay of Whales was into terra incognita. Amundsen named the mountains that rose to block his path the Queen Maud Mountains, after the monarch of his fledgling nation. Other features that he named were Mount Fridtjof Nansen, after his mentor in polar exploration, and Liv Glacier, pictured on page 14, after his nanny and housekeeper. For his route, Amundsen chose instead Axel Heiberg Glacier 20 miles to the south, named for a rich patron, which the party successfully navigated onto the polar plateau in three days.

The Queen Maud Mountains extend from south of Beardmore Glacier to Scott Glacier. Blocky massifs with layered Beacon strata dominate the landscape as far as Axel Heiberg Glacier, but beyond that the sedimentary rocks are largely eroded away.

Near the mouth of Liv Glacier stands a marvelous anomaly, a 1,500-foot horn of pure marble, named the Tusk by a New Zealand geological party that traversed the area in the early 1960s. Overridden by a much thicker Liv Glacier at a time in the past when the ice shelf was grounded and backed up into the mountains to higher elevations, the peak is vertical to overhung in the upstream direction but tapers smoothly at a consistent angle of about 30° downstream.

I first became aware of this beautiful hunk of rock in December 1970, during my first trip to Antarctica. Directly to the north of the Tusk is a shoulder and high ridgeline named Mount Henson, where a helicopter landed my field partner and me after dropping our survival gear at the bottom. We measured and collected 300 meters of stratigraphic section along the ridgeline to the point where the sequence was intruded by granite.

When we finished the section, we hightailed it off the ridge, dropped the samples by our survival gear, and started hiking straight toward the Tusk with time (we thought) to climb and collect it. Much to our chagrin, the helicopter came a couple of hours early and we did not even make it to the foot of the peak.

During the 1974–75 season we worked in the Duncan Mountains, on the far side of Liv Glacier, and traversed it by snowmobile to revisit the area around Mount Henson. This time, no helicopter was going to deny us the Tusk.

People always ask when they see this picture, "Is that you out there?" To which I invariably reply, "No, that's me behind the camera." I was in considerable pain from a severely crooked back as I shuffled up the northern spine of the Tusk. Otherwise, I would have been out there with Phil Colbert, field assistant/mountaineer and my longtime friend, and would have missed the shot. Funny how things turn out sometimes.

Blocky Mount Fridtjof Nansen towers at the skyline with a wall of Basement granite on its lower flank and sedimentary layers of Beacon up to its summit. Liv Glacier flows from the right rear.

To the south and east of Axel Heiberg Glacier the sedimentary layers that characterize the blocky highlands are largely eroded away, and the highest summits are composed of the underlying Basement. The result is a terrain of jagged peaks and narrow ridgelines. I took this photo from slightly below the summit of Mount Griffith (10,154 feet,) looking to the northwest. Amundsen Glacier flows from left to right toward the rear of the image. Beyond that, the flat-topped escarpment, Rawson Plateau, has had all the Beacon sedimentary layers eroded away, exhuming the old erosion surface in the process.

Scott Glacier

Most of the central Scott Glacier area is underlain by granite that erodes into serrated ridgelines and summits. In this view from the summit of Heinous Peak, Watson Escarpment marks the left skyline with the Organ Pipe Peaks and Mount Harkness reaching down to the glacier (compare the photo on page 115). Mount Blackburn rises on the skyline in the middle of the image, while the dark band to the right rear is the La Gorce Mountains. Scott Glacier flows from the right in front of these landmarks.

Midway down the east side of Scott Glacier, a grouping of sublime granite spires named the Gothic Mountains encircles a small glacier known as the Sanctuary. Grizzly Peak (6,630 feet) is the watchtower at the mouth of Sanctuary Glacier where it enters the Scott, flowing past from right to left. Next under cloud is Mount Zanuck (8,285 feet), the highest peak in the ring. Mount Andrews (8,135 feet) and Mount Gerdel (8,267 feet) are the two bright peaks in the left rear.

On the right, the Organ Pipe Peaks complete the circle.

Rock Samples

The Transantarctic Mountains are replete with panoramic scenes—compounded ridgelines and summits and the broad, white spaces in between. But look up close. Those rocks have a story to tell, and pattern and detail to rival that of ice.

Granite

Granite is an intrusive igneous rock formed deep in active mountain belts. Climbers idolize it, its solidness, its marvelous cracks, the shear walls that result when glaciers cut through it. The Scott Glacier area is underlain by a vast tract of this fabled rock. I consider myself more of a mountaineer than a climber. Climbers use ropes and hardware to protect themselves while climbing. I am a klutz with ropes and rarely use them. Plus, as a geologist, I find granite less interesting than almost any other type of rock. Nevertheless, several projects took me to the wonderland of Scott Glacier to climb, collect, and map its granites.

During the 1977–78 season we geologically mapped the last major unvisited area of the Transantarctic Mountains. Except for a single helicopter landing at Mount Webster in 1962, the entire Leverett Glacier quadrangle was terra incognita. We worked from five base camps out from a central put-in site and mapped the entire quadrangle. Snowmobiles were the key, and we logged 50 miles in a day on several occasions. At the end of the season one of the machines registered 600 miles on its odometer.

One of those 50-mile days was to the Berry Peaks in the northeast corner of the quadrangle. We found that each of the peaks was composed of the same, fine-grained granite. Viewed from one of these summits, a steep spire rose to the north punctuated by a smaller shoulder, with the Ross Ice Shelf receding in the distance.

The most spectacular granites in all the Transantarctic Mountains crop out at the circle of peaks known as the Gothic Mountains, midway down the east side of Scott Glacier. The region was first explored in 1933–34 by a three-man geological party, Quin Blackburn, Stuart Paine, and Richard Russell, of Byrd's Second Antarctic Expedition (1933–35). In an enterprise that boasted airplanes and, for the first time, functioning tracked vehicles, these hombres mushed dogs 525 miles to the mountain front from Little America, the base on the Bay of Whales, and then continued another 100 miles up Scott Glacier to Mount Weaver, where they measured more than 1,000 feet of stratigraphic section and collected sixty-seven rock samples. On their return journey to Little America, they lingered for three days in the Gothic Mountains, basking in their grandeur.

Of the geographical features in the Scott Glacier area named by Byrd, only one wasn't for a person, either a member of the expedition or a patron. The single exception is this stunning chain of spires in the Gothic Mountains, which the Blackburn party named the Organ Pipe Peaks. They took photographs and Blackburn published them in his report in the *Geographical Review* in 1937. I read the article in preparation for my first field season, 1970–71, and was dazzled by his grainy, miniature snapshots of the Organ Pipes.

When I saw Blackburn's photos, I dreamed
of standing in his boots staring up in awe at
the central spire. It didn't happen that season,
but I didn't let go of it either.

During the 1980–81 season, I was funded to map the La Gorce Mountains in the southeast quadrant of Scott Glacier. When we completed that work a week before our pull-out date, we conducted a collecting traverse for 45 miles down the east side of Scott Glacier, ending in the Gothic Mountains. From our base camp there we drove snowmobiles around the inner perimeter of the mountains, sampling the tips of each of the spurs that reached down from the peaks.

For purity of form, for grace, for awe, the central spire of the Organ Pipe Peaks is unrivaled. My brother Mugs had wanted to climb it the first time I showed him a picture. All I had wanted to do was to stand in its aura and pay homage. When I brought Mugs on as the party's field assistant/mountaineer, I began to imagine the unthinkable.

With our collecting completed and one

day left before returning to our put-in site in the La Gorce Mountains, Mugs and I took the chance. Mugs had scoped the backside of the peak from camp. It offered a straightforward route up a *couloir* (snow chute) to a shoulder of rock for the first half of the climb. But after that he couldn't tell. "We'll just wander around on the face and see where it leads," he said.

The morning of the climb we ate a big bowl of spaghetti and drove over to the foot of the peak. Mugs kicked all the steps in the snow up to the shoulder and I followed in them. At the shoulder we roped, Mugs led, and we worked our way across, then up the vertical portion of the face that was mostly snow free. This was two pitches or about 300 feet. The stretch where we roped was a series of big cracks. For all but one spot I climbed on Mugs's lead without a problem.

Our climbing route on the backside of the Spectre

The *crux* is the hardest move on a climb, and what follows was the one for me. I was completely exposed standing on a six-inch wide shelf at the foot of a vertical face of granite that rose to above my head and then broke back from a lip. I could grasp the edge and pull myself up to look over it, and what I saw was a face sloping at about 45° away from me that topped in a horizontal crack where I could get my next good handhold. I pulled myself up to my waist and leaned in

over the edge, but the crack was a tantalizing 10 inches out beyond my fingertips. The only way for me to reach the crack would be to pull myself up hard and lunge over the lip in one move. Sitting about 20 feet above me, Mugs tightened the rope. I lunged and missed and stood myself down on the shelf again. Mugs chuckled. He hadn't been telling me what to do the whole climb. And he didn't start then.

Mugs was only an inch taller than me, so I was pretty sure he hadn't done it with a lunge. Then I noticed a knob of rock to the left, about three feet beyond the end of the ledge I was standing on. Leaning toward the knob with handholds above, I swung my body toward the protrusion and grabbed it between my legs. Holding tight with my knees, I was able to reach a handhold in another crack farther to the left and from there I was away.

After about three hundred vertical feet, the face pitched over to an angle of around 60° and Mugs asked, "How are you feelin'? Comfortable?"

"Feelin' good!" was my reply. "It's your call."

Mugs knew my aversion to ropes and smiled. "Well, I would never let a client go up something like this without a rope. But [snort], you're my brother." And then he laughed out loud.

So, we packed the ropes and continued. For the most part it was straightforward kicking steps in steep, snow-filed cracks. The last barrier was a small cornice about 10 feet high and vertical above the rock we were standing on. Mugs got out both of his ice tools and flailed a deep furrow in the loose snow. I followed and like that we were on the top. We hugged each other. There was a round of "Well, all right! All right!" and then we scouted the summit, a broad, relatively flat top, with different panoramas depending on where we were standing. An unusual three-pronged snow drift claimed the northwest foot of the peak.

The brothers Stump atop the Spectre

We ate a lunch of candy bars and gorp, and talked about how proud our parents would be, even if they wouldn't let on to others. It was a family moment. And then we quit the mountain. I rappelled down to the top of the snow chute and Mugs downclimbed where he could, recovering the gear, and rappelling where he couldn't, leaving some slings on the face. From the top of the chute, we glissaded to the bottom and drove back to base camp, one pair of happy campers.

This 1,500-foot face is on a scale with the Empire State Building.

Intrusive Contacts

When granite magma rises, buoyantly intruding the rocks that enclose it, it forces itself into cracks, separating chunks of metamorphic rocks that may sink into the ascending magma. The boundary between solid rock and melt is a fascinating region of the otherworld.

During the 1974–75 season I worked in the Duncan Mountains, a small range between Liv and Axel Heiberg Glaciers adjacent to the Ross Ice Shelf, where I studied a sequence of dark-colored, metamorphic rocks. At several localities, they were intruded by small bodies of light-colored granite, producing striking patterns at the contacts between the two rock types. On this 1,500-foot face, dislodged blocks of metamorphic rock floated in the granite melt before it solidified.

The rock face shown here is about 300 feet across, about the size of a football field. The three, large, dark masses are metamorphic rocks that were intruded by two generations of magma. The first intrusion was the light-gray mass in the middle and left of the image, followed by the lighter-colored veins that finger their way into both of these rocks.

During the 1974–75 season we occupied three base camps, two low ones adjacent to the ice shelf and a high one next to a ridge that connected the Duncan Mountains and a peak named Mount Fairweather (6,120 feet).

In one section of the ridge numerous chunks of metamorphic rock were entrained in a granite intrusion. Their alignment indicates the nearly vertical direction of movement in the magma.

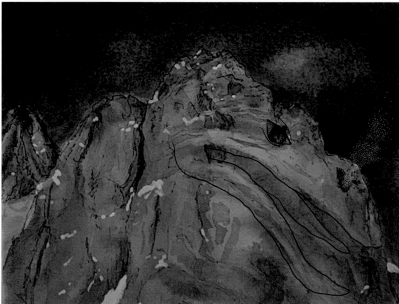

A fire-breather in this land of ice, the dragon surveys its lair.

During the 1987–88 field season in the Scott Glacier area, our second camp was next to Mount Pulitzer (7,070 feet). The massif is elongate with a ridgeline that runs out in opposite directions from the summit. In profile it resembles a saurian creature, such as a dinosaur or dragon. It has legs formed by two spurs that descend from the main ridgeline, and the south end of the ridge is capped with a small peak, shown here, that we informally named the "Dragon's Head." The dragon's mouth is formed by a set of light-colored, arching veins and the eye is a large inclusion of metamorphic rock

Marble

At elevated temperature and pressure some rocks flow as a ductile solid, like glacier ice. One rock that is particularly ductile is marble, metamorphosed limestone. During the 2000–2001 season, our second camp was at the foot of Mount Madison, on the south bank of Byrd Glacier. One day we climbed a thick band of highly deformed marble that snaked its way up a 1,500-foot spur to the summit.

86

Most of the unit was pure white or pale gray, but a succession of dark layers occurred at one horizon revealing that the rock had flowed like taffy. Horizontal distance, ~10 feet.

Vinson Massif

The Ellsworth Mountains are an anomaly. Their geology is different from both the Transantarctic Mountains and the system of mountains in Marie Byrd Land and the Antarctic Peninsula. Isolated in the interior of West Antarctica, they boast the highest elevation on the continent at the Vinson Massif (16,050 feet), and on its western flank an astonishing 11,000 feet of relief. Relief is the difference in elevation between the highest and lowest points on a mountain— or any terrain for that matter. The Grand Canyon has a relief of about 5,000 feet from the South Rim to the bottom. For the Vinson Massif that is more relief than two Grand Canyons!

It was that relief that drew me to the Vinson in 1989–90, when I was funded to climb and collect the massif toward determining its uplift history using a technique called fission-track dating. The field plan was a climber's dream: Find a peak with the greatest relief, climb it to its summit, and on the descent collect 20-pound samples of rock at 100-meter vertical intervals for analysis back in the lab. That it was necessary to reach the summit was because that was the only surveyed spot on the map where we could set our altimeters, and sample elevation was a critical part of the data. This was pre-GPS. The collecting lines and sample sites are shown on the sketch. We dated the samples back at the lab, and the conclusions of the study were that the Vinson Massif had been uplifted 4 kilometers (2.4 miles) or more at a rate of about 200 meters (600 feet) per million years between 141 and 117 million years ago, coincident with the breakup of the supercontinent of Pangea.

The sketch shows the collecting sites (brown dots), collecting lines (white), and the snowmobile routes (blue) that we followed to reach specific rock outcrops. The collections spread over five days. We collected the summit line in two segments, the first day to an elevation of 13,000 feet and the second day the top three samples, including the summit.

Mount Erebus

A plume of vapor bellows from the summit of Mount Erebus, Earth's southernmost active volcano, rising 12,450 feet out of the icebound Ross Sea. Mount Erebus and its inactive twin, Mount Terror (10,597 feet), comprise Ross Island. A narrow, 50-mile-long peninsula of volcanic rock extends south from the main portion of the island, and at its tip are located the US and New Zealand bases, McMurdo Station and Scott Base, respectively.

In the austral summer of 1840–41, a British expedition led by Captain James Clark Ross sailed south along the 170° east meridian with hopes of locating the Magnetic South Pole. Ross never reached the magnetic pole, which was deep in East Antarctica; however, in the course of exploration, he made important discoveries, notably the Transantarctic Mountains and the great ice barrier that would come to be called the Ross Ice Shelf. Ross also discovered Mount Erebus, "emitting smoke and flame in great profusion." Although this sounds to have been a more active phase of eruption than has since been observed, every subsequent group returning to Ross Island has recorded summit activity on the volcano.

In December 1978, I had the opportunity to fly by helicopter from McMurdo to the top of Mount Erebus. The pilot landed us 1,200 feet below the summit on a relatively level spot next to a small monitoring station, one that is still operating to this day. Our party scampered up a path of rock and scree, puffing in the thin air, keen to peer over the edge.

The first ascent of Mount Erebus was by a party from Shackleton's 1907–9 expedition and included four geologists. They had reported a vertical crater wall that dropped 900 feet and ended on a floor of solidified lava with a circular, inner crater, wherein lay the volcanic activity.

As I reached the lip of the volcano, my eyes and lungs were smarting from the heavy sulfur fumes that swirled around us. My heart was also pounding, as much from the anticipation as from the exertion. Here I was at the summit of an active volcano, my first and only this lifetime. The first thing that smacked me in the face was the 900-foot wall on the far side of the crater, just as reported, with horizons that emanated vapor that was both rising out of the crater and freezing at the vents.

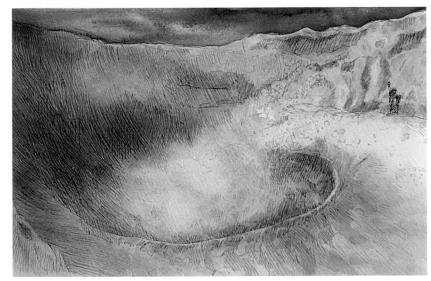

To my left at the end of the crater was the inner crater, belching a cloud of noxious steam. We had picked a good day for our jaunt and were able to see down the inner wall at times. Often if the lava lake is active, clouds of vapor fill the crater and envelope the summit, blocking observation.

The inner crater on Mount Erebus measures approximately 750 feet across and 300 feet deep. The wall at that end of the outer crater is approximately 600 feet high, twice the height of the Statue of Liberty.

We worked our way along the lip of the crater, to the point closest to the inner crater, and waited for the cloud to lift. Several times over a ten-minute interval the cloud pulled back, allowing us a glimpse of the activity. The lake was covered with a thin crust that moved up and down and slowly slid under itself along seams. The lava glowed red at thin places in the crust. Horizontal distance, ~30 feet.

Otherworld meets underworld in the maw of a volcano.

Ice-Cored Moraines

Ice-cored moraines are interesting features of the Transantarctic Mountains. Most valleys are filled up to their ridgelines with glaciers that flow down valleys, merge with other glacial streams, and join one of the outlet glaciers, ultimately ending up in the Ross Ice Shelf or Ross Sea. However, small valleys on the sides of some mountains do not receive enough snow to maintain an active glacier. The result is an accumulation of eroded rock debris from valley walls onto stagnant ice, known as an ice-cored moraine. Motion does occur within the underlying ice, but not enough to flush the system, so lobate ridges of loose rock pile up, tracing the movement of the ice beneath.

This image is of Moraine Canyon on the west side of Nilsen Plateau. The canyon opens into Amundsen Glacier, but its glacier is so diminished that it no longer joins the larger stream.

This ice-cored moraine originates on the north side of an unnamed peak in the upper Scott Glacier area, which is shielded from the prevailing winds and where little snow accumulates. Furthermore, at the height of the austral summer, the sun is high enough to cause melting against the dark rocks. Bartlett Glacier flows from left to right in the background. A stagnant lobe of that glacier noses up to the moraine. Ablation, both melting and evaporation from the solid state, has removed the outer covering of snow, revealing the glacier ice beneath. Faulkner Escarpment bounds Nilsen Plateau in the background.

Gargoyles

In January 2011 a helicopter dropped my party at the crest of the Cobham Range, on the north side of Nimrod Glacier, for a day of mapping. I had been there fifteen years before and climbed to that point. This was a rare time when I had the opportunity to revisit a locality. The ridge that extends to the north of the main range was named Gargoyle Ridge by New Zealand geologists who had mapped it in 1964–65, and I could see why. At the crestline, the rock was eroded into these grotesque, animate forms.

O'Brien Peak

Our put-in and pull-out site for the 1987–88 season in Scott Glacier was about 3 miles south of a nubbin of rock named O'Brien Peak that sits at the edge of the Ross Ice Shelf. I had spent a day mapping there in 1971, and the night before our pull-out I drove over to see if I could spot anything that I remembered.

The sky was overcast and a wind with light, fresh snow was blowing at 30 miles per hour. I drove up the backside of the mountain and looked out across the ice shelf. Meltwater had ponded at the foot of the rock earlier in the season but now was frozen through. The ice shelf was as flat as the sea upon which it floated. The dark horizon was dramatic.

The gray-tan marble at O'Brien Peak was
as I remembered. Icing the rock, gale-force
winds smeared splotches of fresh snow into
cracks as if applied with a pallet knife.

3

THE WIND

In the lifeless world of Antarctica, the wind is an animate force active in human-time—miles per hour, a heartbeat. It may be fierce, it may be calm, it may be steady, it may be restless or fickle or faint. Sometimes it isn't even there at all. It is the bearer of cloud and the deliverer of snow. It can be a fearsome force, roiling through the mountains. In this image, wind hurls clouds across Heinous Peak (top right-center) and Mount Pulitzer (lower left).

Out on the ridgeline, wind challenges
your balance, drains heat from your core,
and nips at the tips of your fingers and nose.

Katabatic Wind

Antarctica is an exception to the rule that Earth's lower atmosphere cools with increasing elevation. Because the ice sheets are so cold, they chill the air above them, creating a continental-scale temperature inversion, with a layer of colder air beneath warmer. Because the chilled air is denser than the air above it, it flows outward under the pull of gravity across the subtly inclined surfaces of the ice sheets. Known as *inversion* or *katabatic winds*, these dense, gravity-driven currents are a stable feature of Antarctica's interior. The average speed of katabatic wind across the East Antarctic Ice Sheet is approximately 10 miles per hour;

however, because the ice sheets are generally steeper around their margins, wind speeds there may increase dramatically, especially down troughs where ice streams enter the ocean and winds funnel through. As katabatic winds drain the base of the atmosphere, the air is replaced by upwardly rising air currents from the Southern Ocean that build and settle over Antarctica.

On the left half of this image, backlit blowing snow traces the outflow of katabatic wind from the East Antarctic Ice Sheet (rear) into the upper reaches of Scott Glacier, as it funnels between Mount Verlautz (8,170 feet, left) and Mount Wyatt (9,613 feet, right).

Polynya

An unusual feature forms at the southern end of Terra Nova Bay, where katabatic winds pour off the East Antarctic Ice Sheet, funnel down Reeves Glacier, and jet into the bay. Once the winter freeze of sea ice has broken into pack, the blast of wind pushes the floes away from the shore creating an area of open water, known as a *polynya*. With the open water driven by both wind and the freezing cold, the surface of the polynya is patterned with swept-winged streams of frazil, the slushy first form of sea ice to freeze. The Terra Nova Bay polynya is a perennial feature, albeit always changing, that I have witnessed numerous times.

In this image, the polynya opens in the lee of Inexpressible Island from wind channeled through Reeves Glacier flowing from the rear around the two, small *nunataks* (islands of rock surrounded by ice).

For scale, Manhattan Island looks small in the expanse of the Drygalski Glacier Tongue.

In this image, the Terra Nova Bay polynya is to the right and rear of the Drygalski Glacier Tongue. Directly behind it is Reeves Glacier. Sea ice has also pulled back at the terminus of the Drygalski, creating a second opening filled with streamers of frazil.

Lenticular Clouds

When conditions are right, inversion winds run up the backside of the Transantarctic Mountains and blow out over a cushion of static air. When the inversion layer encounters peaks, it is forced to rise as it moves forward, producing lenticular cloud formations that trace the passage of the wind.

The necessary condition for the formation of these clouds, or any clouds for that matter, is that the air be saturated in water vapor. When the inversion wind is forced to rise above a peak, its pressure drops because of the increased elevation. At lower pressure, air can hold less water vapor and so the ex-

cess condenses as clouds. Since the air temperature is well below freezing, these clouds are composed of tiny ice crystals rather than droplets of water. Having passed the peaks, the wind descends, and the clouds absorb back into the vapor state.

In this image, lenticular clouds wave above flat-topped Evans Butte. The source of the katabatic winds that produced them is the polar plateau beyond Watson Escarpment, which appears in the left distance with its own katabatic cloud and peeks through the two saddles to the right of Evans Butte.

Lenticular clouds may last for hours or
even days if conditions are ideal. Since the
clouds are bound to the peaks, where the
wind undergoes decreased pressure, they
remain stationary even though the wind is
constantly moving past, but they also move
up and down subtly as the wind varies. A
bank of "lennies" hangs beyond Mount Griffith, the summit in the upper left.

Blowing Snow

When winds move across the glaciers and ice sheets, they pick up ice particles and carry them along as blowing snow, so even if a storm brings no precipitation, near the surface a blinding blizzard may be in progress.

In general, winds of about 20 miles per hour are enough to produce blowing snow a foot or two deep, and by 30 miles per hour a layer is raised well above one's head.

The onset of blowing snow is audible. The soft whisper of wind across the snowfields gains a faint hiss as fine streams of snow, like sand, begin to twist their way through the sastrugi and over the rocks. It takes less energy to keep snow crystals moving than it does to get them moving in the first place, so they sometimes are a portent of stronger winds to come.

And when stronger winds do come, it is wise to hole up in camp and wait them out.

With the added lift of a 35-mile-per-hour
headwind, the C-130 reached gingerly for
the glacier surface. Katabatic winds that
poured over us for most of our stay at the
pickup site had increased considerably that
day. Streamers of blowing snow curled at
the foot of Mount Wyatt (9,613 feet).

Drifts

If the wind carries blowing snow over an outcrop of rock, some of it may drop out in the lee of the obstacle and accumulate downwind as a drift. Drifts are as permanent as the prevailing winds themselves, in places sintered to blue from frigid temperatures and time. Here drifts grow from lows in the ridge at the head of Somero Glacier, draining into Liv Glacier. Mount Schervill (6,545 feet) is the prominent peak in shadow.

Drifts cling to a ridgeline on the Lowry Massif and smear the downwind slope. Byrd Glacier surges in glacier-time from left to right. Beyond that the Britannia Range rises in rock-time, its shoulders laid bare by the elements. In human-time, a seeker stands lost in the scene.

Viewed from the summit, a frigid drift of blue ice extends north from the toe of Mount Stump (8,170 feet), west side of central Scott Glacier. The line of small dots in the upper right are the tents and snowmobiles of our camp.

Wind Scours

When wind blasts into steep rock, a back eddy is formed that scours a trench in the glacier. The same thing happens upwind of one's tent in a snowstorm—a drift builds, always at a certain distance out from the tent. The depressions at glacier margins can be more than a hundred feet deep and may represent serious impediments to reaching particular rock outcrops.

In some cases, snow accumulates at the lip of a wind scour, forming a vertical or overhung wedge known as a *cornice*. Unlike alpine settings where active snowstorms produce cornices that send avalanches into the valleys below, in Antarctica the cornices are solid and stable, seldom chunking off. In this image, cornices circle an outcrop of bedrock near Monte Cassino in northern Victoria Land.

Sastrugi

If you told me I could have one more hour in Antarctica before I died—at a spot of my choice—that a magic carpet would whisk me to wherever I wanted to go, it would not be to one of the sublime vistas where I have stood in awe. It would not be to one of those precious outcrops where I discovered a contact and helped set the clock of geologic time. It would not be to the summit of the Vinson Massif nor of steamy Mount Erebus. Not beneath the face of the awesome Spectre nor in the deep blue interior of a crevasse. It would not be the Adélie penguin rookery at Cape Royds.

If I had one hour more to savor Antarctica, it would be on a névé—a snowfield, circled at a distance by low mountains, snow gracefully rising to narrow ridgelines. A light breeze would nip my nose to remind me of where I was. The midnight sun would be low in the southern sky, casting long shadows and a faint alpine glow. And I would be standing in the midst of a field of the most exquisite sastrugi—wind-carved snow—as far as the eye could see.

How many times did I go out after dinner, tired from the day but still wired, not ready for sleep, and step into that other-world of windswept form, a shifting bas-relief transformed from season to season, from storm to storm? Open to any sort of game that lurked in the field, I would stalk the elusive image, tiptoe up to a promising patch, zooming and framing, and if the image showed itself, "click," I would shoot it.

How many times have I held my breath, focusing, framing the shot just a bit more precisely, zooming slightly in and out, refocusing, checking the corners of the frame again, noticing something different, more zooming, more framing? Thirty seconds might pass, and the low-oxygen alarms would be going off in my lungs. I would slowly squeeze the shutter button for the perfect snap, and "Gasp!"—nothing—realizing in the moment that I hadn't advanced the film since my last shot. It always brought a groan and a chuckle.

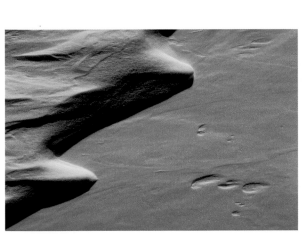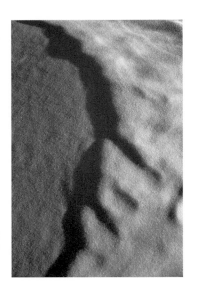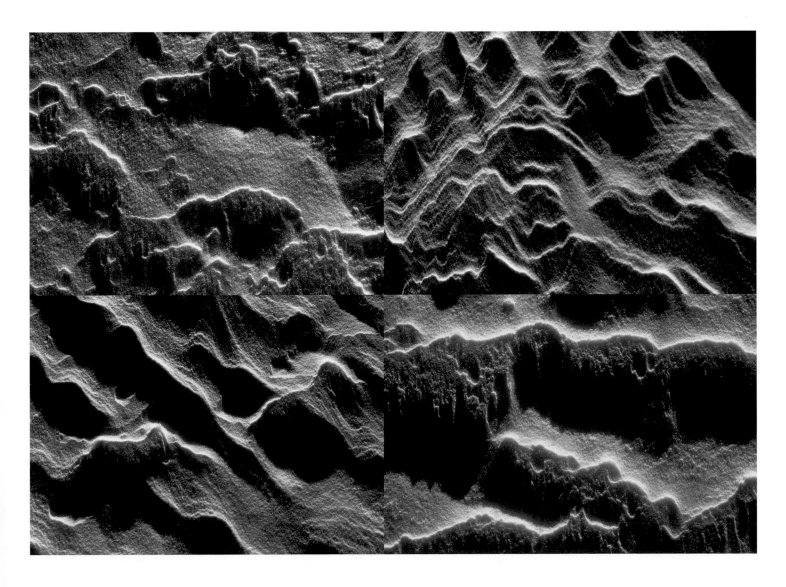

AFTERWORD

Were you ever out in the Great Alone, when the moon
    was awful clear,
And the icy mountains hemmed you in with a silence
    you most could *hear*?

When I was a child, my father would recite "The Shooting of Dan McGrew" by heart. We had a worn copy of the *Collected Poems of Robert Service* and no television, and many were the evenings that he would read to us from it. Among my favorites were "The Ballad of the Ice-worm Cocktail" and "The Ballad of How McPherson Held the Floor." And, of course, "The Cremation of Sam McGee." But "Dan McGrew" was the best. The rootin'-tootin' saloon, the stranger ("in a buckskin shirt that was glazed with dirt"), the melodrama, the gunfire and revenge. But what struck me most, even as a child of seven or eight, was the couplet quoted above. Who knew at the time?

To fly over such terrain as the Transantarctic Mountains is to be transported by flying carpet to an otherworld. I've had many memorable flights by both helicopter and airplane to the length and breadth of the mountains, but the one that stands out above all others was in early November 1978. I was on a reconnaissance flight to upper Scott Glacier to check possible landing sites for our put-in. We planned to fly directly from McMurdo Station to the mouth of Scott Glacier, bank right and track up the glacier. The Gothic Mountains on the east side of the glacier were a fantasy that I had carried from before I ever set foot on the continent.

As we approached Scott Glacier, we could see that weather was brewing. I was in the cockpit of the C-130, kneeling by a large window, first on the right, then on the left side of the plane. The Gothic Mountains were stewing in cloud. I took a series of twelve photos as we flew past the spectacle, several of which have already appeared here (pages 75, 78, 96). This is my favorite. Mounts Andrews and Gerdel glowed as if illuminated from within. Cloud-covered Mount Zanuck, the highest summit in the circle, rose out of the image on the right. Watson Escarpment framed the rear with a mingling of this world and the other.

AFTERWORD

Fifteen seconds later in full sun, the Organ Pipe Peaks with 2,000 feet of sheer relief were end on, casting long shadows. Mount Harkness was on the right. The background disappeared in cloud.

Our first day in the Ellsworth Mountains, November 20, 1989, a blanket of clouds rolled in over the top of the Sentinel Range and plunged down the 7,500-foot western face. Mount Shinn peaked above the mist in the distance.

Crevasses form in a profusion of patterns wherever ice flows faster than its ductile limit. At the head of a glacier, adjacent to bedrock in steep terrain, a single crevasse typically forms. These have rated their own name: *bergschrund*, from the German for "mountain cleft." During the 1974–75 season we operated from three base camps in the Duncan Mountains. The third was high and on the backside of the mountains adjacent to a long ridge connecting to Mount Fairweather (6,120 feet). Big, gaping bergschrunds opened at several places along the ridge where ice began its movement down Somero Glacier, a tributary to the Liv.

One was so welcoming that we were able to walk in, standing up on a path that descended gently into deepening blue. Because the crevasse was open, snow had blown in and collected over time. The deep end of the crevasse was draped with graceful curtains of snow, soft and fragile, barely coherent. The air was dead. Voices sounded far away. The otherworld was at hand. Vertical distance, ~20 feet.

For the listener, who listens in the snow,
And, nothing himself, beholds
Nothing that is not there and the nothing that is.

—WALLACE STEVENS, "The Snow Man"

The nothingness of Antarctica abounds on the fields of white—the great, windswept plains of the ice sheets and the undulating névés that rise into the mountains and blanket their slopes. With nothing in the landscape for scale, no trees, no roads, no buildings, distance is foreshortened. Is that mountain two miles away or twenty? The ever-circling sun gazes down casting ever-changing shadows. I looked out from a camp near Starshot Glacier in January 1979. Between deep shadow and brilliant white the sun's rays grazed the sastrugi veneer of a nearby slope. Lost in cloud the horizon disappeared.

AFTERWORD

Discovering the ice puddle described on page 52 had been transformative. When we returned from our crossing of Liv Glacier and climbing of the Tusk, we worked one more day from a low camp in the Duncan Mountains. When the sun was high, a melt-water stream again flowed along the front of the mountains, and I was on the lookout for other ice puddles. This is what I found.

A Weddell seal made this breather hole in newly formed frazil—the first form of ice to freeze—off the end of Hut Point. Broken chunks of last winter's sea ice wasted in the summer sun.

AFTERWORD

In December 1981, I was doing grab-and-go sampling by helicopter in the Lanterman Range, central northern Victoria Land. We touched down at Carnes Crag at the northern tip of the range. I collected a sample, recorded attitudes, and stepped up to the edge of the crag. An icefall poured down the side of an unnamed peak. Sledgers Glacier flowed from right to left in front of the Bowers Mountains spotted by clouds. It was one of those moments when the landscape framed itself.

AFTERWORD

As a place to work, as a place to go out every day and do the job, as a place to labor, Antarctica levies its demands. Food, shelter, and clothing must be there before the demands of research even come to bear. And for me the demands of research were scripted from a fairytale of yore, a tale of youth and a mountain made of glass and steeds whose hoofs sparked fire. We went out every day (weather permitting) and climbed mountains. Not necessarily to the top, though that was the requirement for some of my later grants, but out along ridgelines, the only exposures of rock to be had, extending the observations of others as well as my own, on some days finding something unexpected, chasing that lead, pushing the hours in perpetual daylight, sleeping long nights,

free-cycling through the seasons. It was a field geologist's dream.

During the 1980–81 field season, we mapped the La Gorce Mountains from two base camps before traversing down the east side of Scott Glacier to the Gothic Mountains. The La Gorce Mountains are a flat-topped massif that rises gently from beneath the ice sheet to the south, and to the north drops 3,000 feet into a valley between two major ridge systems. Because there is insufficient ice in the valley to produce an outward-flowing glacier, stagnant, ice-cored moraines pattern the bottom of the valley. The day we climbed a spur to the summit, the sky was overcast, the scene was somber, the mood contemplative.

How many times did I come to a resting place, plant both feet firmly, pause to catch my breath, and look out—out across expanses of blue and white, into the emptiness of that empty land? Emptiness without context or scale, notes without sound. How many times did I pause and feel close to that fundamental fondo, to the om of it, each its own perspective on the void?

The day before we were to be picked up from our landing site on upper Scott Glacier, December 1978, we were fooling around in some big bergschrunds a mile from camp. Three of us had walked up the crest of a drift, following each other's footprints in search of another crevasse. Scott Borg had lingered and as he turned and shouldered his pack, he cast a long shadow.

Sastrugi on the distant névé trace the path of katabatic winds that flow out of the south and are deflected by the mountain where we stood. At the left rear, Mount Howe, the southernmost outcrop of rock on the planet, stands sentry to the vast beyond.

The night before Mugs and I climbed the central spire of the Organ Pipe Peaks, I had trouble going to sleep. I tossed for a long time in my bag thinking about what the following day might hold, but eventually did drift off. I awoke several hours later, alert but no longer wired. The tent walls were slack, no one down the line was snoring, and an absolute stillness hung on the camp. A soft, very high-pitched tone resonated in my ears, but whether its origin was from within my head or without I wasn't sure. I lay there for ten minutes, acutely aware of the sound of my breath, hearing my every movement in the sleeping bag, trying to hold still and stop breathing, sensing outwardly. The sun was directly to the south, illuminating the yellow world that was the interior of my tent. The ringing in my ears hadn't stopped. Lying there, I felt a presence outside, something beyond what I had seen or sensed that season.

I put on my sunglasses, untied the tunnel entrance to the tent, and pushed my upper body through to the outside. What greeted my eyes were the mountains across Scott Glacier blanketed by those ice-laden lenticular clouds, spread beneath a higher, encroaching weather system. I was spellbound. Although the lennies appeared at first glance to be stationary, in fact they moved slowly and subtly up and down, while the high haze moved in at a perceptible rate.

1
2
3

When I finally broke away and looked to
my left, the peak that I would be climbing
next day was backlit by a naked sun and a
set of undulating lennies that were coming
directly toward me. The ringing in my ears
was still there, and I was starting to be cold.
But a feeling of peace with the morrow dis-
placed my anxiety. I popped back into the
tent to grab my camera, popped back out,
and shot a single frame of the mountains in
cloud, then screwed on my cross filter and
captured the peak that we would later name
the Spectre.

AFTERWORD

The mountains had bestowed on us their beauties, and we adored them with a child's simplicity and revered them with a monk's veneration.

—MAURICE HERZOG, *Annapurna*

The night before we traversed back from the Gothic Mountains to our pickup site, Mugs and I drove over to the prow of the Spectre for a final communion. We chose to hike up to the convergence of the three drifts that we had looked down on the day before. Mugs stopped several times to shoot photos while I walked up the drift ahead of him. From the apex looking back down the line, I shot my fondest image of my brother.

Antarctica has its silences, sound chambers without sound, dead air you can almost hear. During the 1987–88 season we were ready to move camp, but clouds rolled in and turned the snowfields and any route we might choose to flat white. So, we sat out a day as the clouds billowed around the peaks that surrounded camp, with nary a flap of our tents.

Heinous Peak, from which we had collected a 7,500-foot profile several days earlier, rose to the southwest. Clouds all but covered the massif save for the main ridge to the summit. I was out in the cold relishing the "Great Alone."

Wistful could describe my feelings in January 2011 as I flew from McMurdo to Christchurch for the final time. The curtain was about to come down on my forty-year career of unimaginable moments in otherworldly places. And the Transantarctic Mountains were putting on a show. The scene was backlit, the contrast stark. Terra Nova Bay was ice free. In parallel alignment, Cape Washington and the Campbell Glacier Tongue jutted into the northern end of the bay. Low-hanging clouds cloaked the shoulders of Mount Melbourne's volcanic cone (upper right). A tight assemblage of diminishing pack ice patterned the foreground, while high cloud muted the distance and continued northward to blanket northern Victoria Land. This was my last image of that magical land.

Beauty is built into every jot and tittle of creation—into every atomic brick! Beauty soaks reality as water fills a rag. "Hast thou entered into the treasure house of the snow?" the Lord asked of Job from the whirlwind. Beauty is the treasure.

—CHET RAYMO, *Honey from Stone*

During my cruise to the Antarctic Peninsula in December 2014, the southernmost penetration of the ship followed a route through the Lemaire Channel, a dramatic strait bounded by steep walls and peaks. As we circled back around Booth Island, we came to a protected inlet where numerous icebergs had grounded and were in various states of decay. The Zodiacs departed the mother ship, putting quietly into the maze. At one point we sidled up to a small berg, maybe 10 feet long, that was composed of clear ice, not the opaque white-to-blue of normal iceberg ice. This ice had likely formed by freezing to the bottom of a floating ice shelf before breaking free and finding its way into the cove. Sprays of tiny bubbles hung in the clearness, their patterns distorted by the irregular, ablation-pitted surface of the ice. Horizontal distance, 6 feet.

A number of the icebergs in the inlet at Booth Island had melted along deep shelves where waves sloshed in. As I peered into this portal to the otherworld, it peered back. A trail of drop stones led into the depths where a boulder was still stuck in the ceiling.

Think of purest white—the brilliance of all colors—and fathomless blue. Sprinkle in a few dark rocks and the total lack of green and you have Antarctica's minimal pallet.

The last stop of the cruise was Port Lockroy, a British station with a long history and frequent tourist landings. I was over by the adjacent penguin rookery scoping the margins of the bay, savoring those last hours. In the foreshortened world of my telephoto lens, all sense of scale went missing. A shoulder of ice connected back some unknown distance to a wall whose lower slope was in shadow and whose upper slope, skimmed by sunlight, displayed some of Nature's finest plastering.

As the ship departed Antarctica, a host of icebergs drifted beneath a subtle sky as far as the eye could see, subtle except for that cloud in the middle which was anything but. It hung there emphatically, a message from the otherworld. Farewell!

December 28, 2010, my sixty-fourth birthday. I was crossing Beardmore Glacier in a Twin Otter headed back to camp, flying a few hundred feet above the surface. With a bank of cloud filling the lower reaches of the glacier and clear skies up-glacier to the plateau, we were cruising at the interface of *be* and *seem*, of the substantial and the insubstantial, of the real world and the other. The scene was passing by at a steady clip. I was jumping between windows on opposite sides of the plane.

On the near shore of the Beardmore, out the right side of the Otter, clouds lapped into the mountains. I locked on the foreground of the image in my eyepiece, the pattern of spare crevasses, the interplay of light and shadow, of fog and solid ground. I backed off to frame the ridgeline on the right, checked the rear, and clicked the shutter.

AFTERWORD

On the far side of Beardmore Glacier,
some 40 miles away, Wedge Face and a flotil-
la of smaller peaks drifted on a sea of cloud.
As the near ground moved from left to right,
these ships on the horizon appeared to be
sailing from right to left. My eye glued to the
camera, I was lost in the scene. Wedge Face
had reached the gap in front of Barnes Peak,
its summit now framed in cloud. I zoomed
in. Click. And the scene passed—as all do. I
didn't know what I had caught at the time.
Peace to you!

Acknowledgments

In my two previous books on the Transant-arctic Mountains, I acknowledged the many friends and colleagues who have contributed to my research career in Antarctica. Here I wish to acknowledge Ohio State University and the Institute of Polar Studies for the boost that they gave to me and other graduate students at the time by allowing us to submit proposals under our own names to the Office of Polar Programs at the National Science Foundation. In the arcane, privileged world of grantsmanship, faculty usually reserve the right to propose research through a university. Not so Ohio State. The faculty allowed their students to take initiative and then promoted them. Shortly thereafter NSF established the policy that remains in effect to this day—to submit a proposal, a PhD is required—but I slipped in under that wire. I am sure that my grant for the 1974–75 Antarctic field season is what opened the door at Arizona State, where I taught and conducted research until my retirement in 2014.

After retiring I joined ASU's Emeritus College and took a series of creative writing classes that retooled old habits ingrained during decades of writing articles for fellow scientists. The instructors excelled at their tradecraft and brought out the best in my writing. Donis Casey, Pam Davenport, Elizabeth McNeil, Susan Puhlman—thank you for your insights and for keeping me out of trouble those early years after retiring. Many of the assignments for your classes have found their way into this book.

The Emeritus College also has an active writers' group whose members critique one another's work. As the manuscript neared completion, I joined the group and immensely enjoyed their company and their comments. Members at various times included Chris Bayne, Charles Brownson, Winifred Doane, Babs Gordon, Aleksandra Gruzinska, Tony Gully, Dick Jacob, Harvey Smith, and Linda Stryker.

In the year before finishing the book I

started an Instagram account in support of its publication. For this, my first venture into social media, I was ably guided by social media and tech support guru Tracy Valgento.

When my agent, Regina Ryan, left her position as editor in chief at Macmillan Adult Books to become a literary agent, she took with her a passion for developing aspiring authors. I was fortunate to cross paths with her in 2006. She liked my photos but not the voice of the narrative and the lack of good maps. She said so and sent me away with suggestions. A year later, when I returned with an improved manuscript and proper maps, she took me on as a client. And then the serious editing began. Five years later the proof was in the pudding with the publication of *The Roof at the Bottom of the World: Discovering the Transantarctic Mountains* by Yale University Press. Regina has been there at every step of the way in the creation of this book as well. Once again it has been a long haul, this time set against a changing landscape of e-publishing and social media. But again, we've served up a pudding, this time with even more goodies than the first. Dear Regina, as I've said so many times before, I couldn't have done it without you.

This book was a dream that I held from my second Antarctic season, 1974–75. Joseph Calamia made that dream come true. I met Joe in 2011 at the Geological Society of America annual meeting in Minneapolis, where he was manning the Yale University Press book stall. I was scheduled for a meet-and-greet to promote *The Roof at the Bottom of the World*. Few friends or colleagues showed up, but Joe and I hit it off and had a great conversation that lasted the hour. Fast forward twelve years. I was excited to find that Joe was now a senior editor for the University of Chicago Press and thrilled when he said yes to my proposal.

Without a blink, Joe cut out 40 percent of the photos in the original set, rearranged the text, asked simple, clarifying questions, and nudged clearer writing out of me. He wanted a sense of scale for that alien landscape and brought in Marlene Hill Donnelly to provide it through her excellent, playful sketches. Joe, thanks for the opportunity. It has been a pleasure to watch you work and to interact. I have also enjoyed every step of the process working with others on the team, including Matt Lang, editorial associate; Stephen Twilley, production editor; Susan Olin, copy editor; and Nick Lilly, marketing manager. May our labor bear fruit.

Harriet, mother of my children, my biggest fan, the best field assistant I ever had, steadfast, accomplished, funny, unendingly supportive, my adventure buddy to far-flung places and always back to Siena, loved from the depths of my soul, what would this life have been without you?

Appendix

1. Field Personnel during the Author's Antarctic Field Seasons

1970–71, *Queen Maud Mountains*

David Elliot, PI (Principal Investigator) and postdoc, Ohio State
Don Coates, co-PI and postdoc, Ohio State
Jim Collinson, faculty, Ohio State
James Kitching, paleontologist, University of Witwatersrand
John Bergener, MS student, University of Wisconsin
Helmut Ehrenspeck, PhD student, Ohio State
Steve Etter, MS student, Ohio State
Paul Mayewski, PhD student, Ohio State
John Ruben, PhD student, Oregon State
Ed Stump, PhD student, Ohio State
Vaughan Wendland, undergraduate honors student, Ohio State
Phil Colbert, mountaineer

1974–75, *Duncan Mountains*

Ed Stump, PI and PhD student, Ohio State
Charles Corbató, faculty and advisor, Ohio State
Art Browning, MS student, Ohio State
Phil Colbert, mountaineer

1977–78, *Leverett Glacier Area*

Ed Stump, PI and faculty, Arizona State
Greta Heintz, MS student
Pat Lowry, MS student
Phil Colbert, mountaineer

1978–79, Southwest Scott Glacier Area and Area South of Byrd Glacier

Ed Stump, PI and faculty, Arizona State
Michael Sheridan, faculty, Arizona State
Scott Borg, MS student
Patrick Lowry, MS student
Phil Colbert, mountaineer

1980–81, La Gorce Mountains

Ed Stump, PI and faculty, Arizona State
Steve Self, faculty, Arizona State
Jerry Smit, MS student
Philip Colbert, mountaineer
Mugs Stump, mountaineer

1981–82, Northern Victoria Land

Ed Stump, PI and faculty, Arizona State
John Holloway, co-PI and faculty, Arizona State
Scott Borg, PhD student
Kathy Lapham, MS student

1982–83, McMurdo Dry Valleys

Ed Stump, PI and faculty, Arizona State
Harriet Maccracken, field assistant

1985–86, Nimrod Glacier Area

Ed Stump, PI and faculty, Arizona State
Russell Korsch, Australian Geological Survey
David Edgerton, MS student

1987–88, Northwest Scott Glacier Area

Ed Stump, PI and faculty, Arizona State
Paul Fitzgerald, postdoc, Arizona State
Mugs Stump, mountaineer
Lyle Dean, mountaineer

1988–89, Sentinel Range, Ellsworth Mountains

Ed Stump, PI and faculty, Arizona State
Paul Fitzgerald, co-PI and postdoc, Arizona State
Mugs Stump, mountaineer
Rob Hall, mountaineer

2000–2001, Area South of Byrd Glacier

Ed Stump, PI and faculty, Arizona State
Franco Talarico, faculty, University of Siena
Brian Gootee, MS student
"JR" John Roberts, mountaineer

2003–4, Priestley Glacier Area and McMurdo Dry Valleys

Franco Talarico, PI and faculty, University of Siena
Rodolfo Carosi, faculty, University of Torino
Ed Stump, faculty, Arizona State
Patrick Brand, MS student, Arizona State

2010–11, Nimrod Glacier Area

Ed Stump, PI and faculty, Arizona State
Danny Foley, MS student

2. Names of Antarctic Features Proposed by the Author and Approved by the Advisory Committee on Antarctic Names (ACAN), US Board of Geographical Names (BGN)

Mount Analogue 85° 49' 00.0" S, 138° 05' 00.0" W, 3,170 m

Alter Peak 86° 04' 00.0" S, 150° 23' 00.0" W, 1,780 m

Blackwall Glacier 86° 10' 00.0" S, 159° 40' 00.0" W

Mount Colbert 86° 12' 00.0" S, 153° 13' 00.0" W, 2,580 m

Compost Peak 84° 06' 34.0" S, 165° 10' 53.0" E

Contortion Spur 80° 25' 00.0" S, 160° 09' 00.0" E

Mount Corbató 85° 04' 00.0" S, 165° 42' 00.0" W, 1,730 m

Crack Bluff 86° 33' 00.0" S, 158° 38' 00.0" W

Dirtbag Nunatak 85° 32' 00.0" S, 144° 52' 00.0" W, 940 m

Dragon's Lair Névé 85° 51' 00.0" S, 154° 00' 00.0" W

Mount Ehrenspeck 84° 46' 00.0" S, 175° 35' 00.0" W, 2,090 m

Fission Wall 85° 52' 00.0" S, 155° 12' 00.0" W

Forbidden Valley 85° 59' 00.0" S, 154° 00' 00.0" W

Gootee Nunatak 80° 39' 00.0" S, 159° 57' 00.0" E

Gothic Mountains 86° 00' 00.0" S, 150° 00' 00.0" W

Goldstream Peak 86° 41' 00.0" S, 148° 30' 00.0" W, ca. 2,800 m

Grizzly Peak 85° 58' 00.0" S, 151° 22' 00.0" W, 2,200 m

Heinous Peak 85° 59' 00.0" S, 154° 55' 00.0" W, ca. 3,300 m

Hourglass Buttress 86° 40' 00.0" S, 146° 28' 00.0" W

Ivory Tower 85° 28' 00.0" S, 142° 24' 00.0" W, ca. 800 m

Lowry Massif 80° 37' 00.0" S, 158° 12' 00.0" E, ca. 1800 m

Miscast Nunatak 80° 30' 00.0" S, 159° 09' 00.0" E, 910 m

Pallid Peak 84° 37' 00.0" S, 178° 49' 00.0" W, 1,500 m

Roberts Pike 80° 36' 00.0" S, 158° 45' 00.0" E, 1630 m

Sanctuary Glacier 86° 00' 00.0" S, 150° 25' 00.0" W

Sheridan Bluff 86° 53' 00.0" S, 153° 30' 00.0" W

The Spectre 86° 03' 00.0" S, 150° 10' 00.0" W, 2,020 m

Spillway Icefall 85° 01' 00.0" S, 166° 22' 00.0" W

Mount Wendland 84° 42' 00.0" S, 175° 18' 00.0" W, 1,650 m

Wishbone Ridge 84° 56' 00.0" S, 166° 56' 00.0" W

3. Notes about the Author's Photography

Before my first trip to Antarctica in 1970, I didn't own a camera and borrowed one from a friend, a Kodak Pony 135. It had a viewfinder and required a handheld light meter that I seldom used, so at the end of the season I had few photos worth a second look. On the way home that year I stopped at the Navy store in Christchurch, New Zealand, and bought my first single-lens reflex (SLR) camera, a Pentax Spotmatic with a screw mount for attaching its 50 mm lens.

Unable to return to Antarctica due to unprecedented airplane crashes on the continent, I was awarded a travel grant to South Africa in 1972–73 where I continued dissertation research comparing rocks of similar age to my suite in Antarctica. After four months in southern Africa, I took the long way home, backpacking for six months from Cape Town to Cairo to Rawalpindi, where I cashed in my return airfare.

I had set off for South Africa with ten rolls of Agfachrome and a vow that I would make them last till the end of the trip. This was about one photo per day. As a way to ration, I decided that every photo I took should have some geology in it, something that either pertained to my research or would illustrate something for a future geology lecture, if I ever made it as a professor. This parsimony influenced my approach to photography for decades—until I purchased a digital SLR in 2010, a Canon Rebel 2ti with an 18–200 mm zoom lens. Before that, every time I clicked the shutter I would think, "75¢, 75¢." Now the cost per shot never enters my mind, and I shoot way more photos than I should.

The other camera equipment that I took with me on the Africa trip was a 100–300 mm zoom lens, a huge barrel of a thing, that a fellow grad student had offered to me for $25 just before I left. When I returned, I was sold on zoom lenses and fed up with the screw mount. At that time, the 3/4 size 35 mm SLRs were coming onto the market, and I bought a Pentax MX, which I took with me for my second Antarctic season, 1974–75, and every other trip thereafter till I bought my digital. With the quick action of a bayonet mount for switching lenses, a through-the-lens metering system, and everything else manual, I found the MX to be the ideal field camera—lightweight, reliable, and easy to use even when wearing gloves.

Given the kind of research that I did, which involved a lot of strenuous hiking, sometimes on steep terrain, I quickly decided that I didn't have time to take my camera out of my backpack every time that I wanted to shoot a photo. The solution was to wear the camera in its leather case around my neck tucked under a waist buckle on the old NSF-issued parkas. This arrangement worked well, except that I still had to dig into my pack to change lenses between wide-angle and telephoto shots.

Finally, in the mid-1990s, Tamron introduced a 28–200 mm zoom lens that was about three inches long, and I never changed lenses in the field again. I bought a holster case that I strapped firmly on my chest, hung my camera around my neck as always, and kept it snugly in the case ready to be pulled out at a moment's notice. To me this is the ideal camera system for doing fieldwork under extreme conditions. If you are a climber, belly tight against a rock, it doesn't work. But for anything short of this, it frees the hands completely and is ready instantly when the next photo op presents itself.

With the purchase of the Pentax MX for the 1974–75 field season, I discovered the world that exists in the lens of a camera. Low-light conditions are never a problem in Antarctica's summer, so I always shot Kodachrome 25 film. Perhaps the greatest boon to my photo collection was the opportunity to fly long distances over that otherworldly terrain. One of the perks of being the principal investigator on my projects was that on the put-in flights—when the Hercules C-130 would land us at a remote field location—I was welcomed into the cockpit of the Herc to help spot the landing site. I would kneel next to the pilot and gaze out a wide, low window that opened onto splendid terrain that I photographed with abandon.

On the flights to and from Christchurch and McMurdo, I always crowded the porthole in the rear door and stayed there till the last bit of ice was gone or the loadmaster told me to sit down. On helicopters I always had a window seat to direct the pilot where I wanted to go. From the air the world goes by in slow motion unless you are quite close to the ground. The fun for me was to be looking out ahead, spotting some worthy feature, training in on it through the lens, moving in and out with the zoom, waiting until everything aligned—and then I would click.

4. Grants and Publications

The author's grants and publications may be found at https://isearch.asu.edu/profile/10195.